Optimizing Energy Efficiency and Thermal Comfort in Building

Optimizing Energy Efficiency and Thermal Comfort in Building

Guest Editor

Christian Inard

MDPI

Basel • Beijing • Wuhan • Barcelona • Belgrade • Novi Sad • Cluj • Manchester

Guest Editor
Christian Inard
Laboratoire des Sciences de
l'Ingénieur pour
l'Environnement (LASIE,
UMR CNRMS 7356)
La Rochelle Université
La Rochelle
France

Editorial Office
MDPI AG
Grosspeteranlage 5
4052 Basel, Switzerland

This is a reprint of the Special Issue, published open access by the journal *Energies* (ISSN 1996-1073), freely accessible at: https://www.mdpi.com/journal/energies/special_issues/3248XG6R0X.

For citation purposes, cite each article independently as indicated on the article page online and as indicated below:

Lastname, A.A.; Lastname, B.B. Article Title. *Journal Name* **Year**, *Volume Number*, Page Range.

ISBN 978-3-7258-6033-3 (Hbk)
ISBN 978-3-7258-6034-0 (PDF)
https://doi.org/10.3390/books978-3-7258-6034-0

Contents

About the Editor

Christian Inard

Christian Inard has been a Full Professor at La Rochelle University (France) since 1998. He achieved his PhD, entitled "Thermal coupling between heating systems and dwellings", in 1988, and his Accreditation for Supervision of Research in 1996. His research activities are focused on the design of low-energy buildings, indoor thermal comfort and indoor air quality, HVAC systems modelling and control, and urban microclimate modelling and experimentation. He is the author or co-author of more than 150 international papers and communications, and he has supervised 32 PhD candidates. From 2021 to 2025, he served as Vice-rector in charge of research at La Rochelle University.

energies

Review

A Review of Data-Driven Methods in Building Retrofit and Performance Optimization: From the Perspective of Carbon Emission Reductions

Shu-Long Luo [1], Xing Shi [1,2] and Feng Yang [1,2,*]

[1] College of Architecture and Urban Planning, Tongji University, Shanghai 200092, China
[2] Key Laboratory of Ecology and Energy-Saving Study of Dense Habitat, Ministry of Education, Shanghai 200092, China
* Correspondence: yangfeng@tongji.edu.cn; Tel.: +86-21-6598-2420

Abstract: In order to reduce the contribution of the building sector to global greenhouse gas emissions and climate change, it is important to improve the building performance through retrofits from the perspective of carbon emission reductions. Data-driven methods are now widely used in building retrofit research. To better apply data-driven techniques in low-carbon building retrofits, a better understanding is needed of the connections and interactions in optimization objectives and parameters, as well as optimization methods and tools. This paper provides a bibliometric analysis of selected 45 studies, summarizes current research hotspots in the field, discusses gaps to be filled, and proposes potential directions for future work. The results show that (1) the building-performance optimization (BPO) process established through physical simulation methods combines the site, retrofit variables, and carbon-related objectives, and the generated datasets are either directly processed using multi-objective optimization (MOO) algorithms or trained as a surrogate model and iteratively optimized using MOO methods. When a sufficient amount of data is available, data-driven methods can be used to develop mathematical models and use MOO methods for performance optimization from the perspective of building carbon emission reductions. (2) The benefits of retrofits are maximized by holistically taking environmental, economic, and social factors into account; from the perspectives of carbon emissions, costs, thermal comfort, and more, widely adopted strategies include improving the thermal performance of building envelopes, regulating HVAC systems, and utilizing renewable energy. (3) The optimization process based on data-driven methods, such as optimization algorithms and machine learning, apply mathematical models and methods for automatic iterative calculations and screen out the optimal solutions with computer assistance with high efficiency while ensuring accuracy. (4) Only 2.2% and 6.7% of the literature focus on the impacts of human behavior and climate change on building retrofits, respectively. In the future, it is necessary to give further consideration to user behaviors and long-term climate change in the retrofit process, in addition to improving the accuracy of optimization models and exploring the generalization and migration capabilities of surrogate models.

Keywords: carbon emissions reduction; building retrofit; building-performance optimization; data-driven methods; bibliometric analysis

1. Introduction

1.1. Background

Climate change and greenhouse gas emissions (GHGEs) have emerged as critical concerns impacting the sustainable development of global human habitations [1]. Global Carbon Emissions (CEs) are increasing at a rate of about 2% annually [2] and have risen by approximately 10 billion tons over the past two decades [3]. The construction industry contributes ~40% to the global CEs [4], making it a primary driver of global warming [5]. Moreover, with current high urbanization rates and buildings' long lifespans [6], urban construction has entered a phase of stock renewal, necessitating extensive retrofitting of existing buildings now and in the future [7]. Retrofitting buildings represents a practical approach to significantly reducing GHGEs [8] and offering substantial environmental, social, and economic benefits [9], thus plays a pivotal role in advancing long-term sustainable developments [10].

A building retrofit is an intervention to add new materials or elements to existing buildings [11], encompassing methods like upgrading the building envelope, enhancing mechanical systems, and improving operations and management [12]. It serves not only to facelift old buildings but also as a potent means to attain environmental sustainability [13] and building-performance optimization (BPO) [14]. BPO objectives guide the direction of building retrofit strategies [15], and when balancing competing goals such as CEs, energy efficiency, and thermal comfort [16], predictive models are essential for selecting optimal solutions from a range of alternatives [17].

Building-performance prediction models are typically categorized into physical models based on building operational process simulations and mathematical models employing data-driven methods [5]. Physics-based models offer precise and reliable calculations, yet their complexity in input parameters and transient modeling increase computational costs. In contrast, mathematical models provide rapid and efficient performance predictions but require extensive data for model developments and often lack physical interpretations of building-performance parameters [18].

Traditional approaches to retrofitting individual buildings typically rely on architects' knowledge and experience to enumerate and compare options. However, this often faces challenges due to the vast search space and the difficulties in exploring all potential alternatives exhaustively. Advances in computer technology have introduced data-driven methods to the retrofit process, leveraging computer-assisted sampling and iterative calculations for automated evaluations of retrofitting options to derive optimal solutions [19]. Contrasted with traditional methods, this approach overcomes constraints of time and space [20], thereby enhancing the accuracy and efficiency of decision-making processes [21].

Due to the rising complexity of retrofit standards and objectives, surrogate models (also known as metamodels) have garnered significant attention in recent years for their ability to integrate the strengths of both physics-based and mathematical models. These models are trained using a limited set of simulation data as samples. Following parameter adjustments and validations, they can swiftly generate highly accurate results with appropriate input datasets [22], omitting extensive simulations and substantially cutting down calculation times.

Research on building retrofits focusing on Carbon Emission Reductions (CERs) has extensively utilized BPO based on data-driven methods [23–25]. While the optimization procedures in various studies share similarities, differences exist in the methods and tools employed. Thus, it is essential to review existing research, comprehend the development and evolution of the focal points of current research, summarize the characteristics and mechanisms behind the most frequently utilized as well as promising new methods and tools, and provide suggestions and guidance for future research and practice towards the design optimization of CER-based building retrofits.

1.2. Previous Reviews

Previous studies have explored topics related to building retrofits, building CERs, and data-driven methods. Addressing current challenges in building retrofits, scholars have offered theoretical guidance and practical recommendations for implementing retrofit projects, choosing retrofitting tools, and formulating retrofit standards. These efforts encompass perspectives such as sustainable retrofits [14], energy-saving retrofits [26], and decision-making methods [27]. In retrofit projects aimed at CERs, researchers primarily assess the environmental impact across the building's life cycle [11], emphasizing elements like building envelopes, floor plan layouts, and natural ventilation strategies [28] to minimize both Operational Carbon Emissions (OCEs) and Embodied Carbon Emissions (ECEs). Furthermore, data-driven methods play a crucial role in addressing complex issues such as energy predictions and control within buildings [5], contributing to revealing energy-saving potentials. The surrogate models developed through data-driven methods are applied in Building-Performance Simulations (BPSs) and optimization processes to enhance computational efficiency [22].

1.2.1. A Review of Building Retrofit Research

Jagarajan et al. [13] asserted that green buildings can significantly enhance environmental sustainability across design, construction, and maintenance phases, thereby reducing the environmental influence of existing buildings. Their systematic literature review and analysis of green building retrofit initiatives noted a paucity of research concerning critical factors influencing the implementation of retrofit projects, emphasizing the need to focus on the problems faced by stakeholders, such as the lack of financial incentives from the government and the limited availability of green material and technology. Nielsen et al. [14] evaluated decision support tools applicable to building retrofit projects, discussing their application in establishing sustainable development goals, formulating design strategies, and assessing performance. They advocated the use of sustainable standards to provide clearer criteria for screening existing buildings. In a study focused on China, Liu et al. [26] examined obstacles and challenges in building energy-saving retrofit practices and analyzed the evolving national and local incentive measures. Their work contributed to the theoretical foundation of energy-saving policies and standards, proposing a data-driven decision-making system to guide optimal retrofit measure selections. Pombo et al. [29] conducted a review of residential building retrofit research, highlighting the variability in methods for evaluating energy-efficiency measures. They underscored the importance of adopting a life-cycle approach, including methodologies like life cycle assessments (LCAs) and life-cycle costs (LCCs) to identify optimal solutions and assess the retrofit potential of residential buildings. These studies emphasized the multifaceted challenges and opportunities in green building retrofit initiatives, offered insights into enhancing sustainability practices, and supported decision-making framework developments.

1.2.2. A Review of Building CER Retrofit Research

Vilches et al. [11] conducted a comprehensive review on building retrofits and environmental assessments using LCA methods. Their findings highlight that the construction and operational phases are most frequently analyzed, with a primary focus on impact categories such as the global warming potential (GWP) and embodied energy. Mostafavi et al. [28] reviewed 48 studies from 2005 to 2020 investigating energy and carbon performance in high-rise buildings across diverse climates. They identified the significant potential for energy-saving through improvements in envelope designs, floor plan optimizations, and the use of natural ventilation. This study also emphasized strategies to reduce OCEs and ECEs by enhancing thermal performance and integrating recycled materials into building constructions. Li et al. [30] examined sustainable retrofit practices in subtropical high-density cities like Hong Kong, emphasizing the variability of CER outcomes across different building types and environmental conditions. For high-rise and multi-story residential buildings, this study underscored the importance of targeting energy-efficiency improvements in public areas and optimizing building envelopes to meet CER goals. Aghamolaei et al. [31] conducted a research study on decarbonization strategies for university campuses under various climatic conditions. Their research emphasized the critical role of implementing low-carbon and energy-saving measures across campus infrastructures. This review identified key parameters including spatial planning, landscape integration, renewable energy adoption, building envelope enhancement, and sustainable transportation as pivotal for achieving CER goals. Their study also anticipated the Internet of Things (IoTs) and the integration of data-driven technologies to facilitate campus decarbonization efforts through enhanced planning and design processes. These studies contribute to understanding effective strategies and technologies aiming at reducing CEs in buildings and urban environments, providing insights into tailored approaches for different building types and climatic contexts.

1.2.3. A Review of the Application of Data-Driven Methods in Building-Performance Analyses

Wei et al. [5] conducted a thorough review of data-driven methods applied in building energy analyses, encompassing predictive and classification techniques. Their findings underscore the efficacy of these methods in energy load forecasts and modeling, facilitating comprehensive assessments of macro-level energy-saving potentials aligned with consumer demands and the formulation of sustainable urban development strategies. Sun et al. [32] reviewed multi-step energy forecasts using data-driven models. Their comprehensive analysis encompassed feature engineering, model types, and anticipated outcomes, proposing future research directions in predictive capabilities and energy model controls. Roman et al. [16] systematically reviewed the literature on metamodels based on Artificial Neural Networks (ANNs) and BPS. They detailed the methodology for generating ANN-based metamodels, covering stages from sample construction to model training and validation. The authors suggest that for the reproducibility of research results, developers of the metamodel should try to publicize sample databases, which will facilitate more in-depth analyses and wide applications in the future. Westermann et al. [22] explored the application of metamodels in sustainable building design research. They highlighted statistical models as effective surrogates for detailed simulations, offering computational efficiency and mitigating barriers in BPS. Their study synthesized successful applications of metamodels in architectural designs to guide their practical implementation. These studies contribute to advancing methodologies in energy predictions and sustainable analyses through the application of data-driven techniques and metamodeling approaches.

Existing reviews primarily focus on enhancing building energy efficiency or providing decision-making guidance, with limited analyses from the perspective of CERs in building retrofit approaches. Those that discuss CERs tend to compare retrofit strategies from the perspective of different building types or building LCAs, with less discussion on the application of data-driven methods. Reviews of data-driven methods and metamodel applications emphasize energy predictions and sustainable building designs yet rarely integrate these approaches into a comprehensive framework for BPO.

Therefore, based on a systematic review, this article explores the relevance of CERs, building retrofits, and data-driven methods. It introduces research hotspots, compares various data-driven methods applied to enhancing building performance, discusses potentials and challenges in their application in building CER retrofits, and provides suggestions on the selection of appropriate optimization methods from the perspective of CERs for different retrofit conditions and scenarios. It aims to offer an overview of the BPO process using data-driven methods in CER-based retrofits and provide a timely and valuable reference for future research.

1.3. Outline and Structure of This Review

The rest of this article is structured as follows: Section 2 outlines the literature search and selection process (Section 2.1) and methods deployed for bibliometric analyses (Section 2.2) and summarizes the general process of BPO based on data-driven methods (Section 2.3). Section 3 details the methodology for constructing BPO models, covering the simulation method (Section 3.1) and the machine learning method (Section 3.2) and performance indicators and optimization parameters selected during the optimization process. Section 4 summarizes the optimization methods and tools (Section 4.1), as well as the decision-making methods for post-optimization processing (Section 4.2). Building upon these sections, Section 5 discusses the current status and challenges of building retrofit methods (Sections 5.1–5.3) and outlines future research directions (Section 5.4). Section 6 provides a summary of the entire article.

2. Literature Screening and Bibliometric Analysis

2.1. Literature Search and Screening

The literature search and screening process followed the guidelines of the Preferred Reporting Items for Systematic Reviews and Meta-Analyses (PRISMA) [33], ensuring a transparent, consistent, and comprehensive systematic literature review (SLR) [34]. This process comprised four phases: identification, screening, eligibility, and inclusion. This study conducted an extensive literature search across prominent databases in natural sciences and engineering research, including Scopus, Web of Science (WoS), and Science Direct (SD) [35]. To maximize the number of retrieved articles, the initial search avoided terms like "data-driven", which were later applied during literature screening. Keywords encompassed three categories related to "building retrofit", "optimization", and "carbon emission", alongside their synonyms, as outlined in Table 1. Each keyword group was combined using the Boolean operator "AND" during the search process.

Table 1. Literature search keywords.

Term	Keywords
Term 1	"building renovation" OR "building reconstruction" OR "building retrofit *" OR "building refurbishment" OR "building repairment" OR "building restoration" OR "building upgrade" OR "building renewal" OR "building improvement" OR "building reformation"
Term 2	multi-objective OR multi-criteria OR optimization
Term 3	"carbon emission" OR "carbon mitigation" OR "CO_2 emission" OR "CO_2 mitigation" OR "greenhouse gas" OR "global warming" OR "environmental impact" OR "sustainable development"

Note: The wildcard "*" indicates a fuzzy search and is used to replace words with multiple spelling variants, such as "retrofit" and "retrofitting".

Firstly, filtering criteria were applied based on the publication year, document type, and language during the initial search. Secondly, duplicate articles were removed, and the remaining results underwent screening based on title, abstract, and keywords. Articles that aligned with the scope of this paper were then subjected to a full-text review. Finally, all references were meticulously checked, and relevant ones meeting the criteria were included. The screening process adhered to the following criteria:

1. To ensure timeliness, articles published within the last 20 years (2003–2023) were selected. Journal articles, known for their rigorous peer-review process [36], were prioritized for their representation and impact in this field [37]. Conference papers, dissertations, and non-English language publications were excluded, retaining only English-language journal articles. Additionally, to facilitate the focus on research methodologies and processes, review papers were omitted from consideration.
2. The literature reviewed addresses CEs during the operational phase of buildings, covering life-cycle carbon emissions (LCCEs), the GWP, and building environmental effects. Studies focusing solely on energy consumption (EC) without converting it to CE objectives were excluded due to differences in concepts and calculation methodologies. Similarly, the literature exclusively discussing CEs during building construction or other phases was also omitted from consideration.
3. The research must incorporate one or more data-driven methods, such as statistical analyses, optimization algorithms, or machine learning (ML) to optimize building performance. Studies that merely listed and compared retrofit plans without employing these methods were excluded.
4. The literature must address the impacts of envelope retrofit measures on the overall building performance. Studies exclusively focusing on the operations of building mechanical systems, energy structure predictions, and similar topics were excluded. Additionally, articles concentrating solely on specific local building components like curtain wall retrofits and structural seismic performance optimizations were also excluded.

Figure 1 illustrates the literature search and screening process, which was applied to retrieve 611 articles from the database. Following screening and supplementation based on the outlined criteria, 45 articles were selected for a detailed analysis. Based on their content, the relevant studies are summarized in terms of building location and type, optimization objectives and variables, and data-driven methods such as machine learning, sensitivity analyses, optimizations, and decision-making methods, as shown in Tables 2–4.

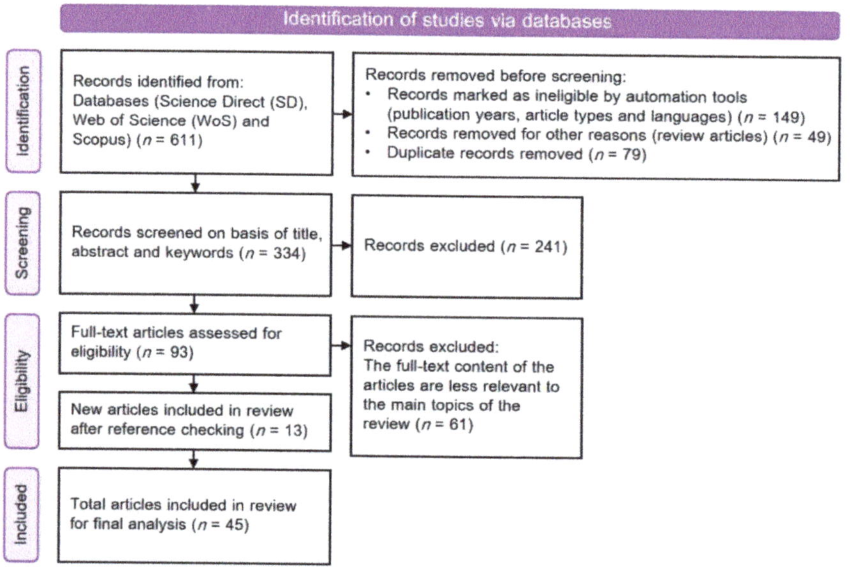

Figure 1. Literature search and screening process.

Table 2. Summary of previous research studies (optimization objectives and variables).

Refs.	Location	Building Type	Optimization Objective	Optimization Variable
[23]	UK	Office building	LCCs, LCEs, and LCCEs	Insulation material area of roof and exterior wall, equipment and energy system, PV panel area, and solar thermal device.
[24]	Switzerland	Residential building	CEs and ACs	U-/R-value of roof, exterior wall, and ground; window type; equipment and energy system; PV system; and solar thermal device.
[25]	Italy	Office building	ECs, TDHs, GCs, and GHGEs	Surface material characteristics of roof and exterior wall, insulation material thickness of roof and exterior wall, window type, equipment and energy system, sunshade component, and PV panel angle and area.
[38]	UK	Office building	LCCEs and OCE	Insulation material type of roof and exterior wall, equipment and energy system, and solar thermal device.
[39]	UK	Office building	LCCs, LCEs, and LCCEs	Insulation material type and area of roof and exterior wall, window-to-wall ratio, equipment and energy system, PV panel area, and solar thermal device.
[40]	China	Shopping mall	OCEs	U/R-value of exterior wall, Glass material, Sunshade component and equipment and energy system.
[41]	Iran	Residential building	CEs and TDHs	Insulation material thickness of roof and exterior wall, insulation material thickness and type of ground, window type, airtightness, and equipment and energy system.
[42]	Canada	Office building	ECs and CEs	Insulation material type of roof, exterior wall, and floor; window type; airtightness; and equipment and energy system.

Table 2. *Cont.*

Refs.	Location	Building Type	Optimization Objective	Optimization Variable
[43]	Finland	Office building	LCCs, RCs, CEs, and TDHs	Insulation material thickness of roof and exterior wall, window type, sunshade component, equipment and energy system, and PV system.
[44]	Iran	Residential building	ECs and the GWP	Insulation material type and thickness of exterior wall and exterior wall type (combination of different materials).
[45]	Canada	Educational building	ECs, LCCs, and LCAs	Type of roof and exterior wall, glass material, airtightness, window opening percentage, and equipment and energy system.
[46]	Korea	Residential building	RCs, LCCs, LCCEs, and CERs	Insulation material type and thickness of exterior wall, window type, sunshade component, and equipment and energy system.
[47]	France	Educational building	ECs, TDHs, RCs, and CEs	Type of roof, floor, ground, and interior and exterior wall; window type, and sunshade component.
[48]	Europe	Residential building	ECs, RCs, OCs, and CEs	Surface material characteristics of roof and exterior wall, window type, sunshade component, sunspace, building form, PV panel angle and area, and solar thermal device.
[49]	Finland	Residential building	ECs, LCCs, and CEs	Insulation material thickness of roof and exterior wall, window type, door material, PV panel area, solar thermal device, and equipment and energy system.
[50]	Iran	Residential building	CEs, WCs, LCCs, and TDHs	Insulation material type and thickness of roof and exterior wall, glass material, filling gas, PV panel area, and equipment and energy system.
[51]	China	Residential building	CEs, TDHs, and GCs	Surface material characteristics, insulation material type and thickness of roof and exterior wall, window type, sunshade component, sunspace, and PV panel angle and area.
[52]	China	Residential building	ECs, RCs, and CERs	Insulation material type of roof, exterior wall and floor, glass material, window-to-wall ratio, and sunspace.
[53]	Estonia	Residential building	GCs, ECs, and LCCEs	Insulation material thickness of exterior wall, surface material characteristics of roof, window type, door material, and building form.
[54]	Korea	Educational building	ECs, CEs, RCs, and TDHs	Type of roof, floor, ground, ceiling, and interior and exterior wall; window type; and equipment and energy system.
[55]	China	Residential building	The GWP, LCCs, and TDHs	Insulation material type and thickness of roof and exterior wall, window type, window-to-wall ratio, and sunshade component.
[56]	China	Residential building	ECs, LCCEs, and LCCs	Insulation material type and thickness of floor and exterior wall, glass material, window-to-wall ratio, sunshade component, and Airtightness.
[57]	UK	Residential building	LCCEs and LCCs	Insulation material thickness, exterior wall type, and window-to-wall ratio.
[58]	Sweden	Residential building	LCEs, LCCEs, and LCCs	Insulation material type and thickness of exterior wall, roof, and ground and window type.
[59]	Switzerland	Residential building	ACs CEs	U-/R-value of roof, floor, and exterior wall; window type; PV panel area; and solar thermal device, and equipment and energy system.
[60]	Canada	Residential building	LCCEs and LCCs	Insulation material type of ceiling and exterior wall, window frame material, door material, airtightness, and equipment and energy system.

Table 2. *Cont.*

Refs.	Location	Building Type	Optimization Objective	Optimization Variable
[61]	UK	Non-domestic building	BER	Type of roof and exterior wall, window type, and equipment and energy system.
[62]	Italy	Residential building	ECs, OCs, RCs, and CEs	Insulation material thickness of roof, floor, and exterior wall; surface material characteristics of roof and exterior wall; PV panel angle and area; glass material; sunshade component; building form; sunspace; and solar thermal device.
[63]	Iran	Residential building	CEs and TDHs	Insulation material thickness of roof, ground, and exterior wall; window type, airtightness, and equipment and energy system.
[64]	Denmark	Residential building	ECs, the GWP, OCs, and RCs	Insulation material type and thickness of interior wall, insulation material type and thickness of roof and exterior wall, surface material characteristics of roof and exterior wall, window frame material, glass material, PV panel area, solar thermal device, and equipment and energy system.
[65]	Italy	Residential building	RCs, OCs, ECs, and CEs	Insulation material thickness of roof, floor, and exterior wall; surface material characteristics of exterior wall; PV panel angle and area; sunshade component, building form, sunspace, and solar thermal device.
[66]	Bosnia and Herzegovina	Residential building	ECs, CEs, and RCs	Insulation material thickness of ceiling and exterior wall, window type, and equipment and energy system.
[67]	China	Office building	ECs, CEs, and TDHs	PV panel angle and area and equipment and energy system.
[68]	Switzerland	Residential building	LCCs and GHGEs	Type of roof and exterior wall, window type, airtightness, PV system, solar thermal device, and equipment and energy system.
[69]	Germany	Residential building	ACs and CEs	Type of roof and exterior wall, window type, PV system, solar thermal device, and equipment and energy system.
[70]	China	Office building	ECs, CEs, and OCs	U-/R-value of roof and exterior wall, window type, and equipment and energy system.
[71]	Greece	Residential building	GHGEs and LCCs	Insulation material thickness of roof, ground, and exterior wall; window type, PV system, solar thermal device, and equipment and energy system.
[72]	China	Office building	LCCEs	Insulation material thickness of roof and exterior wall, surface material characteristics of exterior wall, window type, PV panel area, and equipment and energy system.
[73]	China	Educational building	ECs and LCCEs	Type of roof, floor, and exterior wall; filling gas; building form; insulation material thickness of floor and exterior wall; window frame material; glass material; building form; insulation material thickness of roof; window-to-wall ratio; sunshade component; PV panel area; and equipment and energy system.
[74]	USA	Residential building	GHGEs, WCs, the SQOL, and LCCs	U-/R-value of roof, ceiling and exterior wall, glass material, window-to-wall ratio, and equipment and energy system.
[75]	UK	Residential building	LCCEs and LCCs	Type of roof, floor, ceiling, and interior and exterior wall and window type.
[76]	Canada	Educational building	ECs, LCCs, and LCAs	Type of roof and exterior wall, glass material, window frame material, window-to-wall ratio, airtightness, window opening percentage, and equipment and energy system.

Table 2. *Cont.*

Refs.	Location	Building Type	Optimization Objective	Optimization Variable
[77]	Canada	Office building	ECs, ECEs, and LCCs	Type of roof and exterior wall, glass material, window frame material, window-to-wall ratio, airtightness, sunshade component, and equipment and energy system.
[78]	China	Office building	ECs, CEs, and LCCs	Insulation material type and thickness of roof and exterior wall and window type.
[79]	Switzerland	Residential building	LCCs and LCAs	Insulation material type of ceiling and exterior wall; insulation material thickness of ceiling, floor, and exterior wall; glass material; and window frame material.

Table 3. Summary of previous research studies (machine learning and sensitivity analysis method).

Refs.	Location	Building Type	Machine Learning Method (Accuracy)	Sensitivity Analysis Method
[23]	UK	Office building	-	-
[24]	Switzerland	Residential building	ANN ($R^2 = 0.94$)	-
[25]	Italy	Office building	-	-
[38]	UK	Office building	-	-
[39]	UK	Office building	-	LSA
[40]	China	Shopping mall	-	LSA
[41]	Iran	Residential building	-	GSA (DOE)
[42]	Canada	Office building	-	LSA
[43]	Finland	Office building	-	-
[44]	Iran	Residential building	-	-
[45]	Canada	Educational building	ANN ($MSE_1 = 0.016$ and $MSE_2 = 0.056$)	-
[46]	Korea	Residential building	-	-
[47]	France	Educational building	-	-
[48]	Europe	Residential building	-	-
[49]	Finland	Residential building	-	-
[50]	Iran	Residential building	-	-
[51]	China	Residential building	-	GSA (PCC and SRRC)
[52]	China	Residential building	-	-
[53]	Estonia	Residential building	-	-
[54]	Korea	Educational building	-	-
[55]	China	Residential building	DNN ($R^2 > 0.99$, CV (RMSE) $\leq 1\%$, and NMBE $\leq 0.2\%$)	GSA
[56]	China	Residential building	-	-
[57]	UK	Residential building	-	-
[58]	Sweden	Residential building	-	-
[59]	Switzerland	Residential building	-	-
[60]	Canada	Residential building	-	-
[61]	UK	Non-domestic building	GBRT (RMSE = 1.7%)	LSA

Table 3. *Cont.*

Refs.	Location	Building Type	Machine Learning Method (Accuracy)	Sensitivity Analysis Method
[62]	Italy	Residential building	-	GSA (SRRC)
[63]	Iran	Residential building	-	GSA (DOE)
[64]	Denmark	Residential building	-	-
[65]	Italy	Residential building	-	GSA (SRRC)
[66]	Bosnia and Herzegovina	Residential building	-	-
[67]	China	Office building	-	GSA (SRC)
[68]	Switzerland	Residential building	-	-
[69]	Germany	Residential building	-	-
[70]	China	Office building	-	GSA (Morris)
[71]	Greece	Residential building	-	-
[72]	China	Office building	-	-
[73]	China	Educational building	ANN (MRE = 1.57% R^2 = 0.94)	-
[74]	USA	Residential building	-	-
[75]	UK	Residential building	-	-
[76]	Canada	Educational building	-	-
[77]	Canada	Office building	MVLR and MARSs (MAPE = 0.2–1.8%)	-
[78]	China	Office building	-	-
[79]	Switzerland	Residential building	Gaussian process modelling (Kriging)	GSA (Sobol)

Table 4. Summary of previous research studies (optimization and decision-making method).

Refs.	Location	Building Type	Optimization Method	Decision-Making Method
[23]	UK	Office building	PSO	-
[24]	Switzerland	Residential building	MILP	-
[25]	Italy	Office building	NSGA-II	-
[38]	UK	Office building	PSO	-
[39]	UK	Office building	PSO	-
[40]	China	Shopping mall	Regression	-
[41]	Iran	Residential building	NSGA-II	-
[42]	Canada	Office building	-	-
[43]	Finland	Office building	Pareto-Archive and NSGA-II	-
[44]	Iran	Residential building	Fitness Comparison	-
[45]	Canada	Educational building	NSGA-II	-
[46]	Korea	Residential building	iMOO score	-
[47]	France	Educational building	NSGA-II	-
[48]	Europe	Residential building	aNSGA-II and pNSGA-II	Utopia point
[49]	Finland	Residential building	Pareto-Archive and NSGA-II	-
[50]	Iran	Residential building	prNSGA-III	TOPSIS
[51]	China	Residential building	SPEA2	Utopia point
[52]	China	Residential building	-	Entropy method (Weight of CERs is 30.95%)
[53]	Estonia	Residential building	Regression	-
[54]	Korea	Educational building	NSGA-II/III and MOEA/D	-

Table 4. *Cont.*

Refs.	Location	Building Type	Optimization Method	Decision-Making Method
[55]	China	Residential building	NSGA-II	TOPSIS (Weight of the GWP is 37.29%)
[56]	China	Residential building	NSGA-II	-
[57]	UK	Residential building	NSGA-II	-
[58]	Sweden	Residential building	NSGA-II	-
[59]	Switzerland	Residential building	GA and MILP	-
[60]	Canada	Residential building	NSGA	-
[61]	UK	Non-domestic building	GA	-
[62]	Italy	Residential building	aNSGA-II	Utopia point
[63]	Iran	Residential building	EWSOA	-
[64]	Denmark	Residential building	Omni-Optimizer	Utopia point
[65]	Italy	Residential building	aNSGA-II	Utopia point
[66]	Bosnia and Herzegovina	Residential building	NSGA-III	Desirability function (Weight of CEs is 30%)
[67]	China	Office building	NSGA-II	-
[68]	Switzerland	Residential building	ϵ-constraint	-
[69]	Germany	Residential building	ϵ-constraint	-
[70]	China	Office building	-	-
[71]	Greece	Residential building	MOGA	-
[72]	China	Office building	NOP and MILP	-
[73]	China	Educational building	SEGA	-
[74]	USA	Residential building	GA	-
[75]	UK	Residential building	NSGA-II	-
[76]	Canada	Educational building	NSGA-II	-
[77]	Canada	Office building	-	-
[78]	China	Office building	AHP	-
[79]	Switzerland	Residential building	NSGA-II	-

2.2. Literature Statistics and Bibliometric Analysis

We present a statistical analysis on the quantitative trends, research locations (Figure 2), types of buildings studied, and data-driven methods used (Figure 3) across all reviewed articles. Figure 2 illustrates that due to rapid advancements in computing applications and the growing maturity of building retrofit research [27], significant research contributions have emerged over the past decade. Over time, the volume of publications has steadily increased, underscoring the field's substantial research value and potential for development [80]. Geographically, the majority of studies originate from Europe (n = 21) and Asia (n = 17), with a limited representation of regions such as Africa, South America, and Oceania. Notably, there has been a surge in research output from China since 2022. The location of research and local climate conditions play pivotal roles in building retrofit strategies [81], and the diverse environmental and climatic contexts in Europe and Asia contribute to the breadth of study cases reported in these regions [82].

Figure 3 reveals that residential buildings dominate the literature, reflecting their large scale, high EC, and pollution levels and pressing the need for CER retrofits. Among public buildings, office buildings exhibit more pronounced retrofit effects compared to other commercial buildings due to their stable occupancy intensity and schedule, contrasting with the high population mobility typical of commercial spaces. In addition, each article adopts one or more types of data-driven methods. More than 90% of the reviewed studies use optimization methods for analysis, while the application of machine learning methods is less than 20%, which needs further exploration.

A bibliometric analysis provides insights into the current status and developmental trends within a specific research field, capturing its dynamics and directions [83]. Utilizing scientific meteorological mapping tools allows for a detailed analysis of the literature [84].

Tools such as CiteSpace [85] enable citation burst analyses, visually depicting significant changes and trends in the evolution of a research field [86].

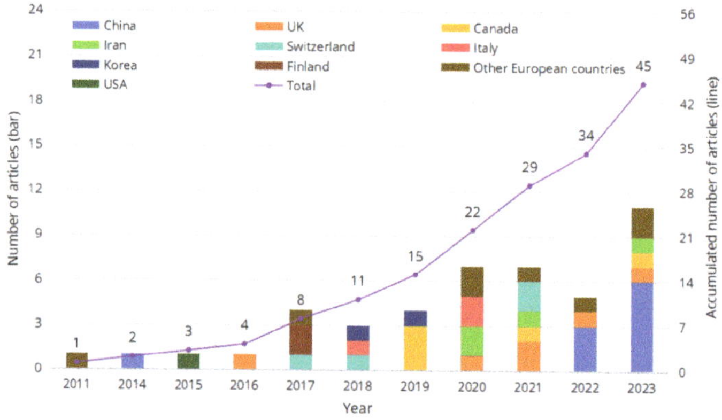

Figure 2. Quantitative trends and geographical distribution of publications.

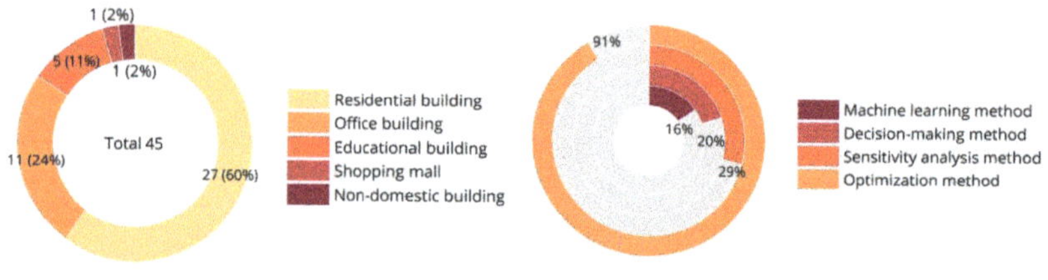

(a) (b)

Figure 3. Building types and data-driven methods of research. (**a**) Building types; (**b**) data-driven methods.

CiteSpace 6.3.R1 was employed to conduct a citation burst analysis (Figure 4). A citation burst analysis highlights trends in disciplinary developments based on the evolution of subject terms [87]. Figure 4 displays the top nine keywords sorted by their initiation year, with the red-line segments indicating periods of significant citation [88]. The early literature extensively discussed "genetic algorithm" (2011–2015), whereas recent attention has been paid to keywords like "embodied carbon" and "machine learning" (2021–2023), indicating a gradual shift towards addressing CEs in building retrofits because of global warming. Moreover, data-driven methods such as "machine learning" began gaining attention in building CER retrofit research around 2021.

Keywords	Year	Strength	Begin	End	2011 - 2023
genetic algorithm	2011	0.82	**2011**	2015	
life cycle cost	2015	0.95	**2015**	2016	
thermal comfort	2017	0.62	**2017**	2020	
energy consumption	2018	1.09	**2018**	2019	
building energy performance	2018	0.74	**2018**	2020	
life cycle assessment	2019	0.78	**2019**	2021	
machine learning	2020	0.71	**2020**	2021	
residential buildings	2020	0.71	**2020**	2021	
embodied carbon	2021	0.9	**2021**	2023	

Figure 4. Top 9 keywords with the strongest citation bursts (the red-line segments indicate periods of significant citation of keywords).

2.3. General Process of BPO Based on Data-Driven Methods

Through a literature review and analysis, it was determined that in the CER retrofit, BPO based on data-driven methods typically adheres to a consistent process. While minor variations exist in the specifics of methods across the literature, the fundamental steps remain the same. These steps are the acquisition of building information, the construction of BPO models, and the screening and decision-making of optimization solutions, as illustrated in Figure 5.

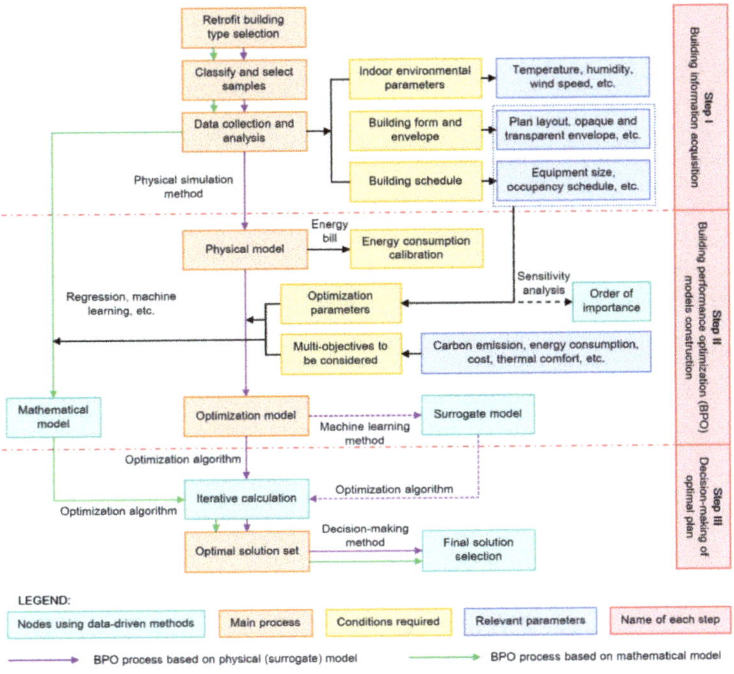

Figure 5. General process of building retrofit and performance optimizations (the dash lines represent the distinction between each step and the dash arrows indicate the optional processes based on computational requirements).

Figure 5 outlines the general process as follows: (1) A suitable retrofit building type is selected, appropriate samples are classified and selected, and pertinent data, including the building form and indoor environmental parameters, etc., are collected through investigation. (2) A physical model is constructed using the above data through simulation methods, and the model accuracy is validated through a comparison with actual data, such as EC records. Optimization parameters and objectives are identified, functional relationships are established between parameters and objectives through simulations and calculations, and an optimization model is developed. Based on the computational efficiency of the model, the choice of whether to use data-driven methods, such as machine learning, is made to further transform it into a surrogate model. When the amount of data is sufficient, a mathematical model can be directly generated based on the collected building information data, combined with optimization objectives and parameters. (3) Optimization algorithms and decision-making methods are integrated to select the retrofit strategy and achieve the optimal solution aligned with retrofit requirements.

3. The Construction BPO Models

The BPO process, shown in Figure 5, can be divided into two categories: (1) the establishment of physics-based BPS protocols (or workflows) based on multiple combinations of site and building energy-carbon-related variables (parameters), and the generated datasets can be either directly processed in MOO algorithms or trained and validated as a surrogate model and then processed with MOO and (2), the development of a mathematical model using regression or a machine learning method, with a sufficient amount of available empirical data, and this is processed with MOO for the optimal solution(s) from the perspective of CERs.

3.1. Optimization Model Based on Physical Simulation Methods

3.1.1. BPS Tools

To construct an optimization model using physical simulation methods, simulation tools are needed to compute building-performance metrics, such as EC calculations based on heat transfer or thermodynamic theories [89]. BPS tools are being made to be more user-friendly by equipping the simulation engines with certain kinds of graphical user interfaces (GUIs) [90]. An analysis of the BPS tools used in the reviewed papers (Table 5) indicated that DesignBuilder [91], OpenStudio [92] (a plugin for SketchUp [93]), and Honeybee [94] (a plugin for Grasshopper [95]) are the primary tools used in simulation research. A comparative study evaluated the strengths and weaknesses of widely used simulation tools and recommended employing a combination of tools and comprehensively assessing data needs and model suitability to enhance modeling accuracies [96].

Table 5. The use of BPS tools in the literature.

Simulation Tool	References
Designbuilder	[25,40,41,44–46,50,63,66,68,73,76–78]
TRNSYS	[39,47,67,71]
SIMEB	[42]
IDA ICE	[43,49,53]
SketchUp—OpenStudio	[48,52,57,62,65,70]
Grasshopper—Honeybee	[51,55,56,58,72]
EnergyPlus	[54,59,74,75]
HOT2000	[60]

3.1.2. Model Calibration

After the physical model is constructed, it is usually necessary to calibrate the model based on monitored weather data and field-measured energy data [97]. The primary calibration process involves adjusting input parameters like heating and cooling set temperatures, indoor occupancy schedules, etc., guided by user experiences and energy consumption records [98].

The model's accuracy is typically assessed by comparing errors between simulated and actual ECs, commonly evaluated against mean bias errors (MBEs) and coefficient of variation of root mean square error CV (RMSE) criteria as defined by ASHRAE Guideline 14 [99], IPMVP [100], and FEMP [101] (Table 6). A smaller error indicates a greater similarity between the physical model and the operational conditions of the reference building [102]. It is worth noting that the above calibration criteria do not take into account the uncertainty of input parameters, which can be further explored through methods such as energy auditing or Bayesian optimization [103].

Table 6. Commonly used error-evaluation indicators.

Evaluation Indicators	Guideline	Monthly Criteria	Hourly Criteria	References
MBE	ASHRAE	±5%	±10%	[46,51,55,56,67]
	IPMVP	-	±5%	
	FEMP	±5%	±10%	
CV (RMSE)	ASHRAE	15%	30%	[46,50,51,55,56,67,77]
	IPMVP	-	20%	
	FEMP	15%	30%	

3.2. Surrogate and Mathematical Models

Unlike physical simulation methods, machine learning methods use mathematical analyses to establish a functional relationship between input and output data in a black-box model. In the BPO process, the surrogate model converted from the optimization model is applied to the optimization process for pre-processing steps [104], which can expedite the generation of samples [105] and decrease computational costs while preserving accuracy [106]. By utilizing historical weather data, energy profiles, and existing building information, the mathematical model generates output values for each retrofit objective [23], which can be effectively applied to large-scale bottom-up building retrofit analyses, such as community and urban scales [24]. However, it demands extensive and accurate historical and statistical data as a foundation [107]. Moreover, predictive variables are typically limited to certain thermo-physical or equipment parameters, constraining the ability to assess the potential for energy savings from retrofits or technological upgrades [108].

Machine learning has emerged as a crucial tool in recent years, leveraging artificial intelligence (AI) to tackle complex tasks and processes in the building sector [109]. It learns from vast datasets and a relatively small number of input features [110] and utilizes learned information for predictions [111]. The process of converting optimization models into surrogate models using machine learning methods typically involves four main steps: data collection and sampling, data pre-processing and model selection, model training and hyperparameter optimization, as well as model validation [112]:

1. The samples of input data are selected using sampling methods, and output data are computed through physical model simulations. Typically, the sample size should range from 10 to 100 times the number of input parameters [113], although this can vary depending on the model's complexity [114].
2. Continuous and discrete variables are distinguished, and both input and output variables are standardized and normalized to ensure a comparability of the data. Typically, the correctly formatted data are divided into training and testing sets in an 80%-to-20% ratio [115], after which an appropriate mathematical model is selected [116].
3. The model is trained, and to prevent overfitting (where the model performs well on the training datasets but struggles with out-of-sample data), hyperparameter optimization is needed to balance variances and biases. When choosing hyperparameters, strategies like a grid search can be employed [117], with cross-validations serving as the scoring method [118].
4. The model is validated, and various metrics are chosen to assess its accuracy. Common evaluation indicators include the mean absolute percent error (MAPE), mean absolute error (MAE), and CV (RMSE) [119], with CV (RMSE) being particularly favored, due to its ability to provide a unitless measurement, which facilitates straightforward comparisons of indicators [120].

In constructing surrogate models, various machine learning methods have been utilized in relevant studies of Table 3, including ANNs [24], gradient boosting regression trees (GBRTs) [61], and multiple adaptive regression splines (MARSs) [77], with ANNs being the predominant approach. An ANN is composed of multiple nodes interconnected in a network [121], including input, hidden, and output layers [122]. This method excels in revealing intricate relationships between input and output data [123], outperforming comparable ML algorithms in processing high-dimensional or nonlinear data, as well as exhibiting a superior prediction accuracy and generalization capabilities [124]. Sharif et al. employed an ANN to develop a surrogate model trained and validated using EC data from DesignBuilder, which was applied thereafter to assess ECs, LCCs, and LCAs in retrofit planning. Their findings demonstrate a strong correlation between surrogate and simulation model outputs, and the computation time can be significantly reduced from 170 h to 150 s [45].

In addition, two articles in the literature chose to explore building retrofitting strategies by establishing mathematical models. Studies [24,61] explored retrofit strategies using 12,806 residential buildings in Zurich and 4900 non-residential buildings in the UK, respectively. The researchers selected a GBRT and an ANN as machine learning methods to directly establish mathematical models, respectively, and the calculation accuracy of the models reached over 90%.

3.3. Optimization Objectives and Parameters

In the process of constructing optimization models, it is necessary to determine optimization objectives and parameters according to the research problem and purpose. The optimization objectives can reflect the building performance before and after the retrofit, while the optimization parameters represent the retrofit means and methods to be taken.

3.3.1. Building-Performance Indicators

The articles reviewed in this paper focus on optimization objectives related to building CEs, encompassing LCCEs and OCEs, among others. Given that OCEs constitute more than 70% of the life cycle [125], some studies calculate CEs specifically during the operational phase rather than across the entire life cycle [30]. However, retrofit measures involve additional materials and components, leading to increased ECEs [126], which must be carefully considered when assessing environmental impacts [29]. Furthermore, certain indicators can gauge the built environment performance and indirectly reflect CEs, such as the GWP, which is used to measure the heat-trapping potential of greenhouse gases in the atmosphere and is expressed as CO_2-eq in calculations [127].

The built environment profoundly impacts personnel comfort, productivity, and well-being [128]. Therefore, the process of building retrofits must integrate environmental, economic, and social criteria to maximize sustainable benefits [129]. Research indicates that ECs, costs, and thermal comfort are the most frequently selected objective functions in optimization studies [130]. Social criteria, such as thermal comfort and indoor environmental quality, enhance the quality of life and satisfaction [131]. Figures 6 and 7 illustrate a heatmap distribution and frequency ranking, showing the co-occurrence of building CE-related objectives with other research goals in the relevant literature of Table 2. It is evident that optimizing building CEs typically considers enhancing ECs, reducing retrofit costs (RCs) and operational costs (OCs), and improving indoor thermal comfort at the same time. There is also significant research focusing on optimizing CEs and costs throughout the building life cycle.

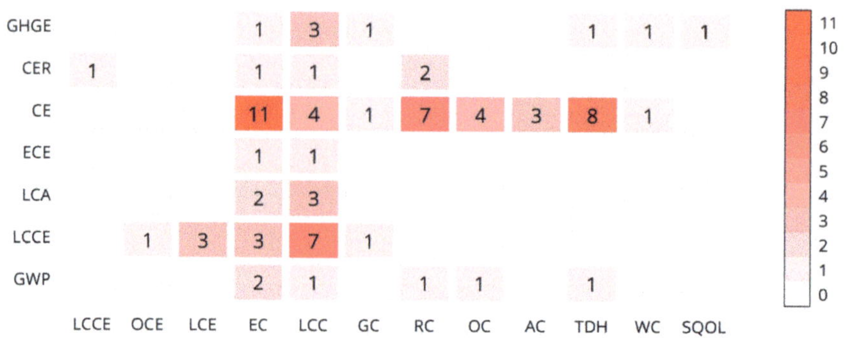

Figure 6. Heatmap distribution of CE objectives with other research goals.

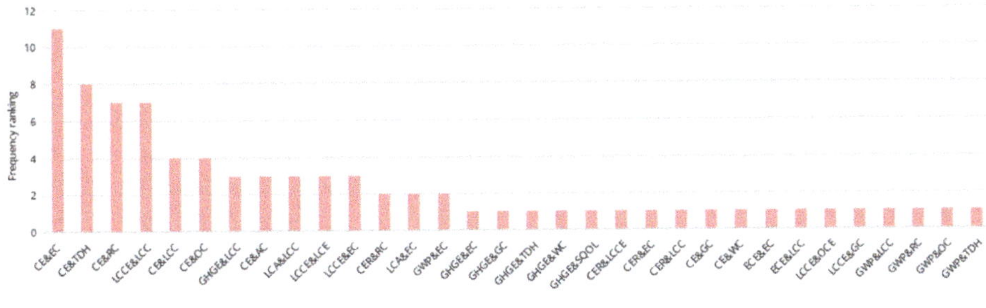

Figure 7. Frequency ranking of CE objectives with other research goals.

3.3.2. Optimization Parameters and Variables

The CER retrofit of buildings necessitates a comprehensive consideration of multiple variables to enhance the overall building performance. The retrofit variables of the literature listed in Table 2 typically fall into two categories: envelope-related variables and other variables (Figure 8). Envelope variables encompass vertical opaque envelopes (interior and exterior walls and doors); horizontal opaque envelopes (ground, floor, ceiling, and roof); as well as transparent envelopes (window-related variables). Others primarily include building forms (such as atria and corridors); sunspaces; sunshade components; building equipment and energy systems (HVAC, lighting, and ventilation systems); and renewable energy systems (like photovoltaic (PV) modules and solar thermal devices). Unlike new building designs, considerations regarding building orientation and form parameters are generally unnecessary in retrofits. Figure 8 illustrates that discussions in the literature often focus on building equipment and energy systems, the window thermal performance, the thickness and type of exterior wall insulations, the thickness of roof insulations, and the area covered by PV panels. Transparent enclosures, although typically smaller in area, exhibit greater heat gains and losses compared to opaque enclosures [132], contributing approximately 20% to 40% of the total [127]. Moreover, rural areas often present greater potential for implementing renewable energy technologies (RETs), facilitated by factors such as accessibility and environmental conditions [133]. Therefore, active and passive measures should be comprehensively utilized during the retrofit to make full use of renewable energy while improving the building performance.

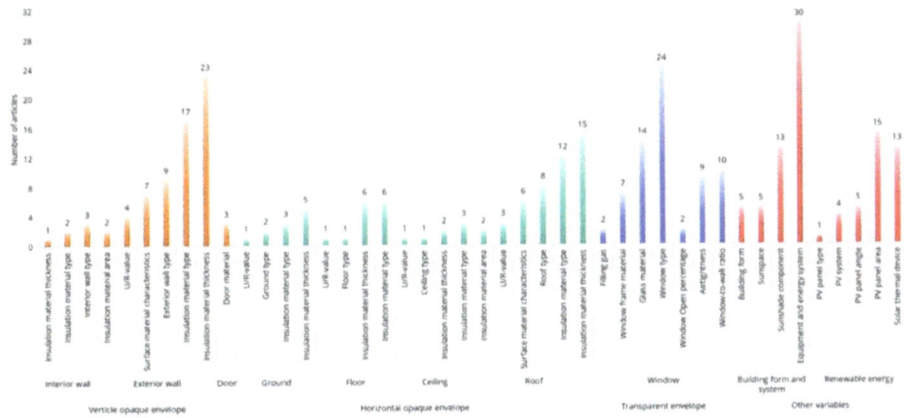

Figure 8. Retrofit variables in studies.

A sensitivity analysis is used to evaluate the correlation between input and output parameters within a model, assessing how changes in inputs influence outputs [134]. It can also be regarded as an optimization search method [135]. For complex scenarios with multiple parameters, a sensitivity analysis can rank the importance of variables [136] and narrow parameter ranges before optimization to simplify model complexity [137]. Before a sensitivity analysis, data sampling is essential to ensure suitable samples, which is often accomplished using Latin hypercube sampling (LHS) [138]. This is a method of stratified random sampling [139] that maintains the representativeness and balance of data while significantly reducing the sample size [140]. Table 7 shows the classification of sensitivity analysis methods and their applications in the literature, with global sensitivity analyses, such as the regression-based standardized regression coefficient (SRC) method [67], the variance-based Sobol method [79], and the screening-based Morris method [70], dominating. The selection of sensitivity analysis methods should consider both model applicability and computational costs comprehensively.

Table 7. Classification and application of sensitivity analysis methods.

Sensitivity Analysis Method		References
LSA		[39,40,42,61]
GSA	Metamodel-based method	[41,63]
	Regression-based method	[51,62,65,67]
	Variance-based method	[55,79]
	Density-based method	[55]
	Screening-based method	[70]

3.3.3. Constraints on Optimization Objectives and Parameters

Retrofitting existing buildings has certain constraints such as geometric shapes, site layouts, functional zones, and structural limitations [141]. These constraints confine the range of feasible variables [73], compelling designers to select retrofit measures within these boundaries. Additionally, constraints often stem from regulatory norms, industry standards, or empirical data. For instance, the total area of solar heater and rooftop PV panels cannot exceed the overall roof area [39]. Design specifications also govern aspects like window-to-wall ratios [55], while budgetary limits constrain retrofit project costs [60]. For CE targets, the ECEs of retrofit measures need to be less than the OCEs that are reduced as a result of an improved building performance. The constraints restrict the optimization range of complex retrofit parameters, prompting the efficient search for solutions that meet requirements.

4. Optimization and Decision-Making Process

4.1. Optimization Process Based on Data-Driven Methods

By reviewing the optimization objectives and parameters, it becomes evident that the process of decarbonizing building retrofits involves numerous variables impacting the building performance. To enhance efficiency, a systematic and effective optimization process is essential [142]. Mathematical methods such as optimization algorithms can be used to sample, screen, and iteratively refine retrofit strategies, so as to efficiently identify optimal solutions. Figure 9 illustrates the uses of various optimization methods in the reviewed articles listed in Table 4, with the non-dominated sorting genetic algorithm II (NSGA-II) being the most frequently employed [143].

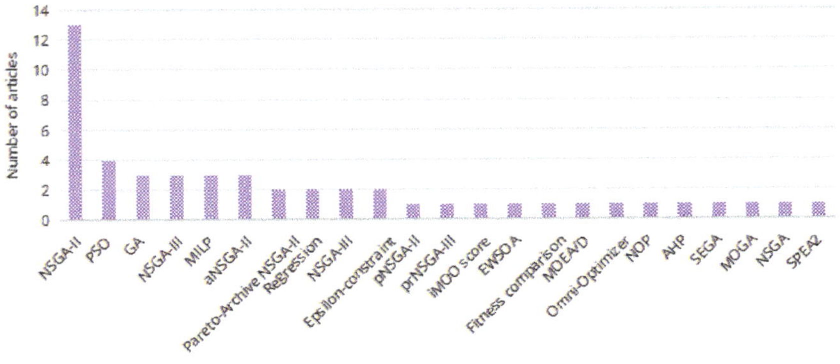

Figure 9. Quantitative distribution of BPO methods.

The exhaustive search method ensures finding the optimal solution but faces an exponentially growing search space with increasing variables, which is often impractical for many real-world optimization problems. A case-by-case analysis through sample screening still relies on enumerated comparisons after sampling and is less utilized in building optimizations [47]. A study employed statistical methods like a regression analysis to model relationships between objective functions and retrofit variables [144]. However, due to the complexity and nonlinearity of BPO problems, the applicability of statistical models may be limited to certain scenarios [145]. To mitigate computational costs and achieve automated iterations and the selection of the optimal solution, researchers have increasingly turned to Evolutionary Algorithms (EAs) for solving optimization problems.

An EA is a population-based random search algorithm [146] known for its faster computational speed, greater accuracy, and enhanced adaptability compared to direct search methods. Including genetic algorithms (GAs), the ant colony optimization (ACO), and the particle swarm optimization (PSO), an EA is effective in navigating the complex search spaces related to building optimization problems [147].

A GA is effective for solving nonlinear and discontinuous problems in building retrofits [148]. It enables a simultaneous search across multiple points in space, thereby reducing the possibility of converging to the local minimum [149]. NSGA-II stands out as one of the most efficient EAs [150], though its performance diminishes with problems involving four or more objectives [151]. NSGA-III [152] offers notable advantages over NSGA-II by enhancing its capability to handle multiple objectives and achieve superior solution distributions [153]. Son et al. examined the optimization of ECs, CEs, RCs, and thermal discomfort hours (TDHs) in public building retrofits, comparing NSGA-II, PSO, multi-objective evolutionary algorithms based on decomposition (MOEA/D), and NSGA-III algorithms. Their findings indicate that NSGA-III exhibits superior diversity and convergence [54]. Furthermore, research has shown that the strength Pareto evolutionary algorithm2 (SPEA2) and NSGA-II perform similarly in solving MOO problems [154], but SPEA2 tends to outperform NSGA-II in high-dimensional objective spaces involving three to four objectives [155].

Similar to GAs, PSO belongs to population-based evolutionary algorithms with meta-heuristic characteristics. However, it does not employ computational processes such as crossovers and mutations [156]. The principle of PSO involves particles searching for optimal solutions by continuously moving through multidimensional search spaces [157].

An optimization analysis necessitates numerous evaluations to achieve near-optimal solutions, leading to relatively extended processing times. Hence, the performance of optimization algorithms plays a crucial role in the efficiency of the optimization process, prompting the selection of suitable methods tailored to specific research requirements [90].

Comprehensive optimization tools integrate multiple algorithms, harmonizing the optimization process. Table 8 illustrates the application of optimization tools in the literature and reveals that Python and MATLAB are the predominant general programming tools utilized in research.

Table 8. The use of optimization tools in the literature.

Optimization Tool	References	Optimization Tool	References
MATLAB	[25,50,72,74]	Python	[48,56,58,60,62,65,71,73,77]
SPSS	[40]	Octopus	[51]
jEPlus + EA	[41,57,67]	JESS + JEA	[63]
MOBO	[43,49,64]	CPLEX	[68]
Excel-VBA	[46]	Gurobi	[69]
MultiOpt	[47]	PLOOTO	[75]

Python is a high-level interpreted programming language. Its ease of learning and vast library of tools have popularized its use in data science and programming [158]. In the context of BPO, Python serves as an application platform for data-driven methods, making the implementation of optimization algorithms more convenient and feasible. For instance, Shadram et al. [58] utilized the pymoo package in Python to implement the NSGA-II algorithm and focused on optimizing the life-cycle energy consumption (LCE), LCCE, and LCC of a conceptualized apartment in various locations in Sweden through retrofit strategies. This study integrated performance simulation and optimization processes using Python and EnergyPlus extension plugins within Grasshopper, establishing a complete and fast optimization process.

MATLAB [159] is a versatile tool for algorithm development, data visualization, and program interaction. The MATLAB Optimization Toolbox facilitates collaborations with other programs, making it an ideal numerical environment for optimization tasks that rely on third-party simulation programs. However, the programming language demands of MATLAB and the complexity of integrating simulation programs can pose challenges for architects. Specialized optimization platforms, like jEPlus + EA [160], can integrate energy simulation programs effectively. Despite offering fewer optimization algorithms, they excel in combining simulation tools and optimization methods with less programming demands, thus enhancing accessibility for architects.

Figure 10 illustrates the relationship among the simulation tools, optimization algorithms, and optimization tools as utilized in the literature. This figure highlights the broad applicability and interaction of tools and methods such as DesignBuilder, NSGA-II, and Python, facilitating diverse applications in research.

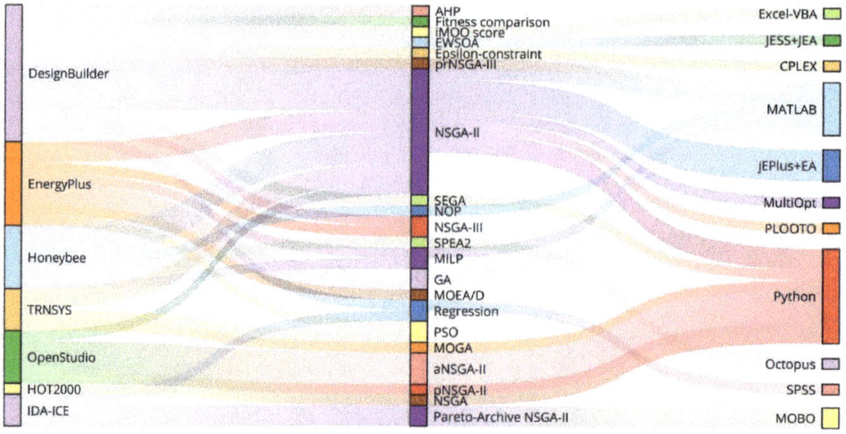

Figure 10. Correspondence among simulation tools, optimization algorithms, and tools.

4.2. Solution Set Evaluation and Decision-Making Methods

For MOO problems, due to conflicting objectives constrained by each other, the resultant solutions often form a set rather than a single solution, known as the non-dominated or Pareto optimal solution set. The spatial distribution of this set is referred to as the Pareto front. Figure 11a,b illustrate typical examples of the Pareto front for two and three objectives, respectively [161]. Throughout the optimization process, each generation of solution sets requires an evaluation based on metrics such as convergence, uniformity, and diversity [162], with assessment indicators including the inverted generational distance (IGD) [163] and the hypervolume index (HI) [164], among others. The quality of the solution set improves as it approaches the true Pareto frontier, exhibiting better uniformity and a wider distribution.

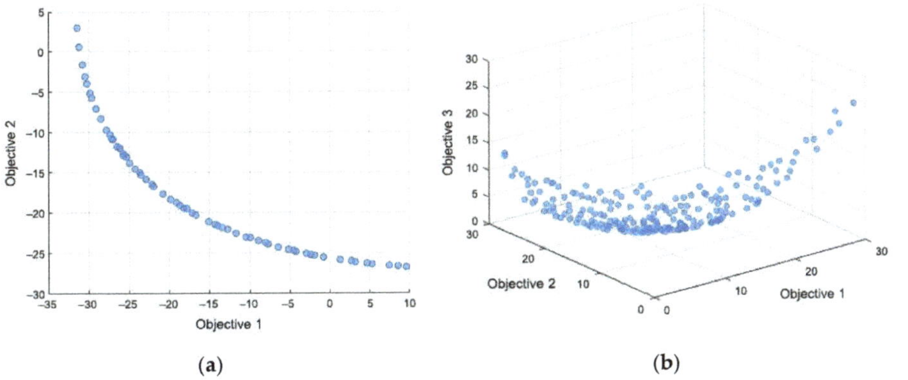

Figure 11. Typical examples of the Pareto front [161]. (**a**) Two-dimensional Pareto front; (**b**) three-dimensional Pareto front (the blue circles represent the solutions to the MOO problem).

Each solution within the Pareto front represents a non-dominated outcome, necessitating post-processing decision-making methods to identify the optimal solution. The technique for order preference by similarity to an ideal solution (TOPSIS) [165] and the Utopia-point method are two commonly employed approaches. TOPSIS ranks solutions by evaluating their distances to both positive and negative ideal solutions, accommodating different weights assigned to objectives based on decision-makers' preferences to derive optimal solutions. Song et al. used the entropy-based method to assign weights to indicators and calculated the information utility value of each indicator based on information entropy, which avoids the influence of human factors compared with subjective empowerment methods. According to the entropy-based method, the GWP, LCCs, and TDHs were assigned weights of 37.29%, 34.17%, and 28.54%, respectively, and were applied to the TOPSIS calculation. The GWP of the final optimized solution can be reduced by 2720 kg/m^2 (about 61%) [55].

The Utopia point serves as an ideal reference within the search space, representing the optimal solutions across all objective functions. Solutions on the Pareto front are assessed by their proximity to this point, with the closest solution indicating the optimal compromise when all objectives are equally weighted. Figure 12a,b depict examples of the TOPSIS [166] and Utopia point [167] methods in 2D coordinates.

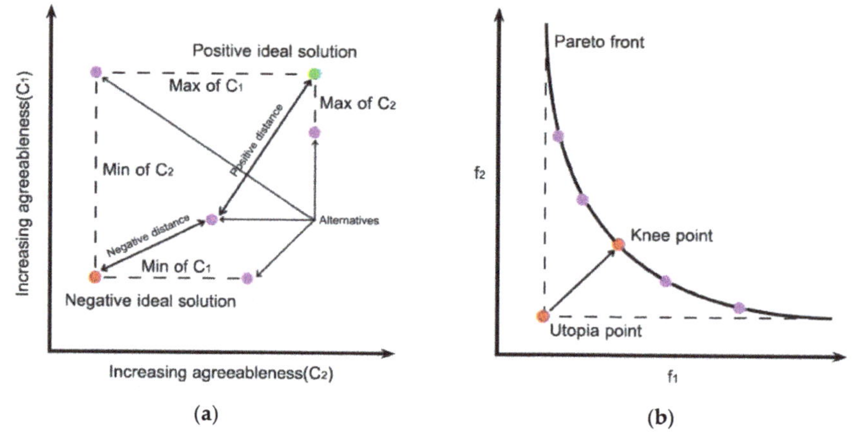

Figure 12. Examples of TOPSIS and Utopia-point methods in 2D coordinates. (**a**) TOPSIS method [166]; (**b**) Utopia-point method [167] (the purple circles represent the solutions to the MOO problem).

5. Discussions

5.1. The Research Status Quo

The increasing prevalence of relevant research in recent years may be attributed to a deeper understanding and increased interest in the significance of CERs within the building construction industry. Advances in science and technology and the widespread adoption of computing have facilitated the continuous evolution of BPS technologies and mathematical algorithms, thereby rendering related research more achievable.

Most research has predominantly been conducted in Europe. Achieving the CER target by 2050 necessitates the extensive decarbonization retrofitting of large-scale existing buildings [168]. Moreover, with the fixed form and orientation of buildings slated for retrofitting, emphasis can be placed on optimizing the envelope and systems, thereby facilitating the application of optimization techniques [21].

Building CER projects typically involve multiple objectives and variables. Integrating BPS with data-driven methods such as optimization algorithms and ML can streamline the process, minimizing redundant tasks and facilitating a quicker and more accurate identification of optimal solutions.

5.2. Optimization and Surrogate Models

Using physical simulation methods to establish an optimization model can accurately reflect the relationship between building parameters and performance and evaluate the performance improvement potential, but the simulation time is relatively long. The use of surrogate models significantly enhances computational speed, and there exist variations in the selection and training of mathematical models across studies. For instance, the number of nodes in the hidden layer of an ANN varies depending on task objectives [169]. Likewise, the division ratio of the datasets into training and testing sets lacks uniformity, and the result reliability may hinge on the training and testing data used in surrogate model developments [135]. Therefore, the choice of model and computational approach should be determined by each task's unique characteristics, such as the data volume and prediction goals [170].

Additionally, addressing issues such as model generalizations and transferability is needed to foster the widespread adoption of surrogate models. Transfer learning methods offer a viable approach to applying data acquired from previous tasks to similar or related tasks [171], which is particularly effective when the training dataset for the targeted task is not large [172]. The successful application of this approach depends on the similarity of data sources and the feasibility for predicting the building performance [173]. Similarly, in energy prediction research for large-scale buildings, transfer learning can solve problems like data scarcity and time-consuming data collection, enhancing the accuracy of data-driven models [174]. A study used transfer learning methods to apply evaluation models pre-trained using Swedish building datasets to predict the performance of buildings in China, which can effectively support building retrofit decisions in data-scarce regions [175].

The clarification and classification of optimization objectives and parameters can provide researchers with experience and paradigms for retrofits and serve as a reference for sensitivity analyses and variable screening outcomes. While focusing on CERs, the relevant studies also take into account optimization goals in terms of building ECs, thermal comfort, and costs [25]. A study also found that cost parameters are particularly common in MOO problems of private housing retrofits, and householders are more willing to carry out CER retrofits on the premise of economic benefits. Therefore, cost savings are as important as CERs, or even more critical [176]. Thus, achieving a balance among multiple performance indicators is essential in retrofitting efforts.

Retrofit strategies typically consider enhancing the performance of building envelopes and adjusting parameters of energy systems. Currently, there is a heightened focus on utilizing renewable energy sources, particularly PV and solar thermal systems. Additionally, emerging technologies and materials, such as switchable blinds [177], PV walls [178], electrochromic windows [179], phase change materials (PCMs), and cool paints [180], among others, are increasingly being employed to enhance the building performance. The vertical greening system (VGS) on roofs and walls are also noteworthy for their potential to mitigate GHGEs, urban heat islands, and noise pollution [181].

5.3. Optimization Methods and Tools

In building CER problems, the choice of optimization methods varies depending on specific objectives and parameters, necessitating tailored selections based on the actual problem [182]. Furthermore, it is crucial to assess the computational efficiency and stability of these methods. While prioritizing computational speed, it is essential to strike a balance between model simplicity and accuracy [97].

The utilization of general optimization platforms like MATLAB necessitates proficiency in a complex programming language, and the constant switching between modeling and optimization environments also poses an inconvenience. Hence, there is a need to integrate programming languages into toolkits that allow for direct applications. The Grasshopper-based Octopus tool [183] integrates optimization algorithms with building parametric modeling and is widely adopted for current BPS and optimization [184–186]. Nevertheless, further exploration is required, particularly in its application to building retrofit processes aimed at achieving CER goals. In addition, regarding the assessment of the solution set quality, a study compared the convergence and diversity using metrics such as IGD
and HI [187].

In recent years, scholars have started to focus on integrating optimization algorithms with Building Information Modeling (BIM) to enhance decision-making abilities in building retrofitting [188]. As data-driven methods advance, there is a corresponding need to further develop tools that integrate these methods with simulation software, enabling timely and efficient applications in BPO. Additionally, establishing a database using existing building performance datasets [189] and implementing a decision support system based on data-driven methods can assist decision-makers in selecting optimal retrofit strategies based on real-world conditions [26].

5.4. Future Work

Uncertainties stemming from user behaviors and climate change present formidable challenges in building retrofitting [8]. Factors such as user preferences [190], comfort requirements [191], and energy-saving awareness exert a substantial influence on the building performance, making user behaviors one of the primary contributors to the disparity between

BPS and actual operation data [192] and a pivotal factor shaping retrofit strategies [193]. For instance, Li et al. [194] integrated three user-behavior patterns—adjusting sunshade components, managing lighting, and ventilating by opening windows—into the optimization of sunshade devices. Their findings revealed how adaptive user behaviors significantly impact room illumination and the simulation of energy loads, thereby influencing the layout and sizing of sunshade panels. Only one of the reviewed studies discussed the impacts of user behaviors on building CEs [70]. This study focused on public buildings in the hot summer and cold winter region of China and implemented a time-of-use pricing strategy to regulate occupants' energy-use behaviors, including the temperature set point of air conditioners and the usage time of electrical appliances. This strategy can reduce users' ECs and CEs by 26.4% and energy expenditure by 26.1%. Current simulation studies often rely on fixed schedules, overlooking active user adjustments and the variability and spontaneity of behaviors. Future studies should incorporate more precise and comprehensive behavioral models into BPS frameworks [142].

A study developed a climate model based on atmospheric CO_2 concentration changes, projecting that the Earth's surface annual average temperature will rise by 1.5 °C by 2050 and by 2–4 °C by 2080 [195]. Given the long lifespan of buildings, climate change should be factored into energy-saving analyses [196] to ensure the applicability and efficacy of retrofit strategies [197]. Data-driven approaches can be used to generate future weather data and establish dynamic climate models.

The building performance is closely correlated with the outdoor environment [198]. Factors such as community greening, urban heat islands, and global warming alter microclimates, influencing the regional building performance [199]. Therefore, using uniform meteorological parameters for predicting building ECs can lead to inaccuracies [200], highlighting the need to integrate microclimate models into energy-efficiency predictions. Currently, optimization research for building CER retrofits has expanded beyond individual buildings to include local outdoor environments, exploring the effects of tree shading and the PV coverage of building façades on the life-cycle GWP of buildings and outdoor thermal comfort [201]. As such, there is still great potential for research on decarbonization retrofits at the scale of building groups or blocks.

6. Conclusions

This paper provides a review of the application of data-driven methods in building retrofits and performance optimization from the perspective of CERs and also discusses recent trends and advancements in various studies. A comparison of modeling methods, optimization variables, objectives, and decision-making methods across 45 relevant papers revealed the following findings:

1. There are usually two workflows to optimize the building performance. One is the workflow of the optimization of the physical simulation (model surrogate) performance: Using the combined input of a building site and energy-carbon-related retrofit variables, a BPO process based on a physical simulation is established. The generated datasets can be either iteratively processed with optimization algorithms directly or trained as a surrogate model, validated, and then processed using the MOO method. The other is the workflow of mathematical modeling-optimization analyses: with sufficient actual field-measured empirical data available, data-driven methods, such as regression or machine learning, are used to develop mathematical models, and multiple objectives are comprehensively optimized from the perspective of building CERs.
2. A building retrofit aims to maximize its benefits by integrating environmental, economic, and social considerations. Therefore, alongside CE objectives, factors like costs and thermal comfort should also be taken into account. There are 27 relevant studies in Table 2 related to the comprehensive optimization of three or more objectives, accounting for 60% of the total. Discussions on retrofit parameters should extensively cover aspects such as the thermal performance of the building envelope, building equipment and energy systems, and the utilization of renewable energy sources.

3. Data-driven methods applied in optimization enable the sampling, screening, and iterative refinement of retrofit plans using computational tools, facilitating the determination of optimal solutions. The advancement and deployment of surrogate models make simplified mathematical calculations replace complex physical simulations, which further enhance optimization efficiency while ensuring accuracy.
4. In the reviewed studies, only 2.2% (1 article) and 6.7% (3 articles) of the total focus on the impacts of human behaviors and climate change on building retrofits, respectively. Future research should delve deeper into the application of data-driven methods in building CER retrofits and BPO, considering user behaviors and variations in retrofit conditions amid long-term climate change scenarios. In addition, more work is needed to improve the accuracy of surrogate models and enhance generalizations and transfer capabilities.

Within the broader context of global warming and building stock renewals, building decarbonizing retrofitting and performance enhancements have emerged as focal points of research. This review highlights the capability of integrating data-driven methods with BPS to screen and determine retrofit plans, showing its significant applicational value in building CERs. It will also make a greater contribution to the CER targets of the building construction industry in the future.

Author Contributions: Conceptualization, S.-L.L. and F.Y.; methodology, S.-L.L., X.S. and F.Y.; software, S.-L.L.; formal analysis, S.-L.L., X.S. and F.Y.; investigation, S.-L.L.; resources, F.Y. and X.S.; data curation, S.-L.L.; writing—original draft preparation, S.-L.L.; writing—review and editing, F.Y. and X.S.; visualization, S.-L.L.; supervision, F.Y.; project administration, F.Y.; funding acquisition, F.Y. All authors have read and agreed to the published version of the manuscript.

Funding: This research was funded by the National Natural Science Foundation of China, grant number 52338004.

Data Availability Statement: No new data were created in this study.

Conflicts of Interest: The authors declare no conflicts of interest.

Abbreviations

AC	Annual cost
ACO	Ant colony optimization
AHP	Analytic hierarchy process
ANN	Artificial neural network
ASHRAE	American Society of Heating, Refrigerating, and Air-Conditioning Engineers
BPO	Building performance optimization
BPS	Building performance simulation
CEs	Carbon emissions
CER	Carbon emission reduction
CV(RMSE)	Coefficient of variation of root mean square error
DNN	Deep neural network
DOEs	Design of experiments
EA	Evolutionary algorithm
EC	Energy consumption
ECEs	Embodied carbon emissions
EWSOA	Enhanced water strider optimization algorithm
FEMP	Federal Energy Management Program
GA	Genetic algorithm
GBRT	Gradient boosting regression tree
GC	Global cost
GHGEs	Greenhouse gas emissions
GSA	Global sensitivity analysis
GWP	Global warming potential
IPMVP	International Performance Measurement and Verification Protocol

LCA	Life-cycle assessment
LCC	Life-cycle cost
LCCE	Life-cycle carbon emission
LCE	Life-cycle energy consumption
LSA	Local sensitivity analysis
MAE	Mean absolute error
MAPE	Mean absolute percent error
MARSs	Multiple adaptive regression splines
MBE	Mean bias error
MILP	Mixed-integer linear programming
MOEA/D	Multi-objective evolutionary algorithm based on decomposition
MOGA	Multi-objective genetic algorithm
MOO	Multi-objective optimization
MVLR	Multi-variate linear regression
NOP	Nonlinear optimization programming
NSGA-II/III	Non-dominated sorting genetic algorithm II/III
a/pNSGA-II	Active/passive archive NSGA-II
prNSGA-III	NSGA-III algorithm augmented by parallel computing structure and result-saving archive
OC	Operational cost
OCES	Operational carbon emissions
PRISMA	Preferred reporting items for systemic reviews and meta-analyses
PSO	Particle swarm optimization
RC	Retrofit cost
SEGA	Strengthen elitist genetic algorithm
SPEA2	Strength Pareto evolutionary algorithm2
SQOL	Social quality of life
TDHS	Thermal discomfort hours
WC	Water consumption

References

1. Gan, V.J.L.; Deng, M.; Tse, K.T.; Chan, C.M.; Lo, I.M.C.; Cheng, J.C.P. Holistic BIM Framework for Sustainable Low Carbon Design of High-Rise Buildings. *J. Clean. Prod.* **2018**, *195*, 1091–1104. [CrossRef]
2. Qian, D.; Li, Y.; Niu, F.; O'Neill, Z. Nationwide Savings Analysis of Energy Conservation Measures in Buildings. *Energy Convers. Manag.* **2019**, *188*, 1–18. [CrossRef]
3. CO_2 and Greenhouse Gas Emissions. Available online: https://ourworldindata.org/co2-and-greenhouse-gas-emissions (accessed on 5 March 2024).
4. Mata, É.; Korpal, A.K.; Cheng, S.H.; Jiménez Navarro, J.P.; Filippidou, F.; Reyna, J.; Wang, R. A Map of Roadmaps for Zero and Low Energy and Carbon Buildings Worldwide. *Environ. Res. Lett.* **2020**, *15*, 113003. [CrossRef]
5. Wei, Y.; Zhang, X.; Shi, Y.; Xia, L.; Pan, S.; Wu, J.; Han, M.; Zhao, X. A Review of Data-Driven Approaches for Prediction and Classification of Building Energy Consumption. *Renew. Sustain. Energy Rev.* **2018**, *82*, 1027–1047. [CrossRef]
6. Love, P.; Arthur Bullen, P. Toward the Sustainable Adaptation of Existing Facilities. *Facilities* **2009**, *27*, 357–367. [CrossRef]
7. Deb, C.; Schlueter, A. Review of Data-Driven Energy Modelling Techniques for Building Retrofit. *Renew. Sustain. Energy Rev.* **2021**, *144*, 110990. [CrossRef]
8. Ma, Z.; Cooper, P.; Daly, D.; Ledo, L. Existing Building Retrofits: Methodology and State-of-the-Art. *Energy Build.* **2012**, *55*, 889–902. [CrossRef]
9. Jensen, P.A.; Maslesa, E.; Berg, J.B.; Thuesen, C. 10 Questions Concerning Sustainable Building Renovation. *Build. Environ.* **2018**, *143*, 130–137. [CrossRef]
10. Huang, J.; Wang, S.; Teng, F.; Feng, W. Thermal Performance Optimization of Envelope in the Energy-Saving Renovation of Existing Residential Buildings. *Energy Build.* **2021**, *247*, 111103. [CrossRef]
11. Vilches, A.; Garcia-Martinez, A.; Sanchez-Montañes, B. Life Cycle Assessment (LCA) of Building Refurbishment: A Literature Review. *Energy Build.* **2017**, *135*, 286–301. [CrossRef]
12. Ji, Y.; Xu, P.; Ye, Y. HVAC Terminal Hourly End-Use Disaggregation in Commercial Buildings with Fourier Series Model. *Energy Build.* **2015**, *97*, 33–46. [CrossRef]
13. Jagarajan, R.; Abdullah Mohd Asmoni, M.N.; Mohammed, A.H.; Jaafar, M.N.; Lee Yim Mei, J.; Baba, M. Green Retrofitting—A Review of Current Status, Implementations and Challenges. *Renew. Sustain. Energy Rev.* **2017**, *67*, 1360–1368. [CrossRef]
14. Nielsen, A.N.; Jensen, R.L.; Larsen, T.S.; Nissen, S.B. Early Stage Decision Support for Sustainable Building Renovation—A Review. *Build. Environ.* **2016**, *103*, 165–181. [CrossRef]
15. De Wilde, P. Ten Questions Concerning Building Performance Analysis. *Build. Environ.* **2019**, *153*, 110–117. [CrossRef]

16. Roman, N.D.; Bre, F.; Fachinotti, V.D.; Lamberts, R. Application and Characterization of Metamodels Based on Artificial Neural Networks for Building Performance Simulation: A Systematic Review. *Energy Build.* **2020**, *217*, 109972. [CrossRef]
17. Wu, Z.; Xia, X.; Wang, B. Improving Building Energy Efficiency by Multiobjective Neighborhood Field Optimization. *Energy Build.* **2015**, *87*, 45–56. [CrossRef]
18. Hong, T.; Langevin, J.; Sun, K. Building Simulation: Ten Challenges. *Build. Simul.* **2018**, *11*, 871–898. [CrossRef]
19. Attia, S.; Hamdy, M.; O'Brien, W.; Carlucci, S. Assessing Gaps and Needs for Integrating Building Performance Optimization Tools in Net Zero Energy Buildings Design. *Energy Build.* **2013**, *60*, 110–124. [CrossRef]
20. Di Biccari, C.; Calcerano, F.; D'Uffizi, F.; Esposito, A.; Campari, M.; Gigliarelli, E. Building Information Modeling and Building Performance Simulation Interoperability: State-of-the-Art and Trends in Current Literature. *Adv. Eng. Inform.* **2022**, *54*, 101753. [CrossRef]
21. Shi, X.; Tian, Z.; Chen, W.; Si, B.; Jin, X. A Review on Building Energy Efficient Design Optimization Rom the Perspective of Architects. *Renew. Sustain. Energy Rev.* **2016**, *65*, 872–884. [CrossRef]
22. Westermann, P.; Evins, R. Surrogate Modelling for Sustainable Building Design—A Review. *Energy Build.* **2019**, *198*, 170–186. [CrossRef]
23. Luo, X.J.; Oyedele, L.O. A Data-Driven Life-Cycle Optimisation Approach for Building Retrofitting: A Comprehensive Assessment on Economy, Energy and Environment. *J. Build. Eng.* **2021**, *43*, 102934. [CrossRef]
24. Thrampoulidis, E.; Mavromatidis, G.; Lucchi, A.; Orehounig, K. A Machine Learning-Based Surrogate Model to Approximate Optimal Building Retrofit Solutions. *Appl. Energy* **2021**, *281*, 116024. [CrossRef]
25. Ascione, F.; Bianco, N.; Mauro, G.; Napolitano, D.; Vanoli, G. A Multi-Criteria Approach to Achieve Constrained Cost-Optimal Energy Retrofits of Buildings by Mitigating Climate Change and Urban Overheating. *Climate* **2018**, *6*, 37. [CrossRef]
26. Liu, Z.; Yu, C.; Qian, Q.K.; Huang, R.; You, K.; Visscher, H.; Zhang, G. Incentive Initiatives on Energy-Efficient Renovation of Existing Buildings towards Carbon–Neutral Blueprints in China: Advancements, Challenges and Prospects. *Energy Build.* **2023**, *296*, 113343. [CrossRef]
27. Li, Y.; Du, H.; Kumaraswamy, S.B. Case-Based Reasoning Approach for Decision-Making in Building Retrofit: A Review. *Build. Environ.* **2024**, *248*, 111030. [CrossRef]
28. Mostafavi, F.; Tahsildoost, M.; Zomorodian, Z. Energy Efficiency and Carbon Emission in High-Rise Buildings: A Review (2005–2020). *Build. Environ.* **2021**, *206*, 108329. [CrossRef]
29. Pombo, O.; Rivela, B.; Neila, J. The Challenge of Sustainable Building Renovation: Assessment of Current Criteria and Future Outlook. *J. Clean. Prod.* **2016**, *123*, 88–100. [CrossRef]
30. Li, J.; Ng, S.T.; Skitmore, M. Review of Low-Carbon Refurbishment Solutions for Residential Buildings with Particular Reference to Multi-Story Buildings in Hong Kong. *Renew. Sustain. Energy Rev.* **2017**, *73*, 393–407. [CrossRef]
31. Aghamolaei, R.; Fallahpour, M. Strategies towards Reducing Carbon Emission in University Campuses: A Comprehensive Review of Both Global and Local Scales. *J. Build. Eng.* **2023**, *76*, 107183. [CrossRef]
32. Sun, Y.; Haghighat, F.; Fung, B.C.M. A Review of The-State-of-the-Art in Data-Driven Approaches for Building Energy Prediction. *Energy Build.* **2020**, *221*, 110022. [CrossRef]
33. Page, M.J.; McKenzie, J.E.; Bossuyt, P.M.; Boutron, I.; Hoffmann, T.C.; Mulrow, C.D.; Shamseer, L.; Tetzlaff, J.M.; Akl, E.A.; Brennan, S.E.; et al. The PRISMA 2020 Statement: An Updated Guideline for Reporting Systematic Reviews. *Syst. Rev.* **2021**, *10*, 89. [CrossRef] [PubMed]
34. Son, T.H.; Weedon, Z.; Yigitcanlar, T.; Sanchez, T.; Corchado, J.M.; Mehmood, R. Algorithmic Urban Planning for Smart and Sustainable Development: Systematic Review of the Literature. *Sustain. Cities Soc.* **2023**, *94*, 104562. [CrossRef]
35. Mongeon, P.; Paul-Hus, A. The Journal Coverage of Web of Science and Scopus: A Comparative Analysis. *Scientometrics* **2016**, *106*, 213–228. [CrossRef]
36. Olawumi, T.O.; Chan, D.W.M. A Scientometric Review of Global Research on Sustainability and Sustainable Development. *J. Clean. Prod.* **2018**, *183*, 231–250. [CrossRef]
37. Santos, R.; Costa, A.A.; Grilo, A. Bibliometric Analysis and Review of Building Information Modelling Literature Published between 2005 and 2015. *Autom. Constr.* **2017**, *80*, 118–136. [CrossRef]
38. Luo, X.J. An Integrated Passive and Active Retrofitting Approach toward Minimum Whole-Life Carbon Footprint. *Energy Build.* **2023**, *295*, 113337. [CrossRef]
39. Luo, X.J.; Oyedele, L.O. Assessment and Optimisation of Life Cycle Environment, Economy and Energy for Building Retrofitting. *Energy Sustain. Dev.* **2021**, *65*, 77–100. [CrossRef]
40. Liang, Y.; Pan, Y.; Yuan, X.; Yang, Y.; Fu, L.; Li, J.; Sun, T.; Huang, Z.; Kosonen, R. Assessment of Operational Carbon Emission Reduction of Energy Conservation Measures for Commercial Buildings: Model Development. *Energy Build.* **2022**, *268*, 112189. [CrossRef]
41. Aghamolaei, R.; Ghaani, M.R. Balancing the Impacts of Energy Efficiency Strategies on Comfort Quality of Interior Places: Application of Optimization Algorithms in Domestic Housing. *J. Build. Eng.* **2020**, *29*, 101174. [CrossRef]
42. Charles, A.; Maref, W.; Ouellet-Plamondon, C.M. Case Study of the Upgrade of an Existing Office Building for Low Energy Consumption and Low Carbon Emissions. *Energy Build.* **2019**, *183*, 151–160. [CrossRef]
43. Niemelä, T.; Levy, K.; Kosonen, R.; Jokisalo, J. Cost-Optimal Renovation Solutions to Maximize Environmental Performance, Indoor Thermal Conditions and Productivity of Office Buildings in Cold Climate. *Sustain. Cities Soc.* **2017**, *32*, 417–434. [CrossRef]

44. Amani, N.; Kiaee, E. Developing a Two-Criteria Framework to Rank Thermal Insulation Materials in Nearly Zero Energy Buildings Using Multi-Objective Optimization Approach. *J. Clean. Prod.* **2020**, *276*, 122592. [CrossRef]
45. Sharif, S.A.; Hammad, A. Developing Surrogate ANN for Selecting Near-Optimal Building Energy Renovation Methods Considering Energy Consumption, LCC and LCA. *J. Build. Eng.* **2019**, *25*, 100790. [CrossRef]
46. Jeong, K.; Hong, T.; Kim, J.; Cho, K. Development of a Multi-Objective Optimization Model for Determining the Optimal CO_2 Emissions Reduction Strategies for a Multi-Family Housing Complex. *Renew. Sustain. Energy Rev.* **2019**, *110*, 118–131. [CrossRef]
47. Chantrelle, F.P.; Lahmidi, H.; Keilholz, W.; Mankibi, M.E.; Michel, P. Development of a Multicriteria Tool for Optimizing the Renovation of Buildings. *Appl. Energy* **2011**, *88*, 1386–1394. [CrossRef]
48. Salata, F.; Ciancio, V.; Dell'Olmo, J.; Golasi, I.; Palusci, O.; Coppi, M. Effects of Local Conditions on the Multi-Variable and Multi-Objective Energy Optimization of Residential Buildings Using Genetic Algorithms. *Appl. Energy* **2020**, *260*, 114289. [CrossRef]
49. Niemelä, T.; Kosonen, R.; Jokisalo, J. Energy Performance and Environmental Impact Analysis of Cost-Optimal Renovation Solutions of Large Panel Apartment Buildings in Finland. *Sustain. Cities Soc.* **2017**, *32*, 9–30. [CrossRef]
50. Mostafazadeh, F.; Eirdmousa, S.J.; Tavakolan, M. Energy, Economic and Comfort Optimization of Building Retrofits Considering Climate Change: A Simulation-Based NSGA-III Approach. *Energy Build.* **2023**, *280*, 112721. [CrossRef]
51. Gao, Y.; Luo, S.; Jiang, J.; Rong, Y. Environmental-Thermal-Economic Performance Trade-off for Rural Residence Retrofitting in the Beijing–Tianjin–Hebei Region, Northern China: A Multi-Objective Optimisation Framework under Different Scenarios. *Energy Build.* **2023**, *286*, 112910. [CrossRef]
52. Li, Q.; Hu, H.; Ma, L.; Wang, Z.; Arıcı, M.; Li, D.; Luo, D.; Jia, J.; Jiang, W.; Qi, H. Evaluation of Energy-Saving Retrofits for Sunspace of Rural Residential Buildings Based on Orthogonal Experiment and Entropy Weight Method. *Energy Sustain. Dev.* **2022**, *70*, 569–580. [CrossRef]
53. Kertsmik, K.-A.; Kuusk, K.; Lylykangas, K.; Kalamees, T. Evaluation of Renovation Strategies: Cost-Optimal, CO_2e Optimal, or Total Energy Optimal? *Energy Build.* **2023**, *287*, 112995. [CrossRef]
54. Son, H.; Kim, C. Evolutionary Many-Objective Optimization for Retrofit Planning in Public Buildings: A Comparative Study. *J. Clean. Prod.* **2018**, *190*, 403–410. [CrossRef]
55. Song, J.; Wang, W.; Ni, P.; Zheng, H.; Zhang, Z.; Zhou, Y. Framework on Low-Carbon Retrofit of Rural Residential Buildings in Arid Areas of Northwest China: A Case Study of Turpan Residential Buildings. *Build. Simul.* **2023**, *16*, 279–297. [CrossRef]
56. Liu, S.; Wang, Y.; Liu, X.; Yang, L.; Zhang, Y.; He, J. How Does Future Climatic Uncertainty Affect Multi-Objective Building Energy Retrofit Decisions? Evidence from Residential Buildings in Subtropical Hong Kong. *Sustain. Cities Soc.* **2023**, *92*, 104482. [CrossRef]
57. Schwartz, Y.; Raslan, R.; Mumovic, D. Implementing Multi Objective Genetic Algorithm for Life Cycle Carbon Footprint and Life Cycle Cost Minimisation: A Building Refurbishment Case Study. *Energy* **2016**, *97*, 58–68. [CrossRef]
58. Shadram, F.; Mukkavaara, J. Improving Life Cycle Sustainability and Profitability of Buildings through Optimization: A Case Study. *Buildings* **2022**, *12*, 497. [CrossRef]
59. Miglani, S.; Orehounig, K.; Carmeliet, J. Integrating a Thermal Model of Ground Source Heat Pumps and Solar Regeneration within Building Energy System Optimization. *Appl. Energy* **2018**, *218*, 78–94. [CrossRef]
60. Zhang, H.; Hewage, K.; Prabatha, T.; Sadiq, R. Life Cycle Thinking-Based Energy Retrofits Evaluation Framework for Canadian Residences: A Pareto Optimization Approach. *Build. Environ.* **2021**, *204*, 108115. [CrossRef]
61. Seyedzadeh, S.; Pour Rahimian, F.; Oliver, S.; Rodriguez, S.; Glesk, I. Machine Learning Modelling for Predicting Non-Domestic Buildings Energy Performance: A Model to Support Deep Energy Retrofit Decision-Making. *Appl. Energy* **2020**, *279*, 115908. [CrossRef]
62. Ciardiello, A.; Rosso, F.; Dell'Olmo, J.; Ciancio, V.; Ferrero, M.; Salata, F. Multi-Objective Approach to the Optimization of Shape and Envelope in Building Energy Design. *Appl. Energy* **2020**, *280*, 115984. [CrossRef]
63. Liu, B.; Pouramini, S. Multi-Objective Optimization for Thermal Comfort Enhancement and Greenhouse Gas Emission Reduction in Residential Buildings Applying Retrofitting Measures by an Enhanced Water Strider Optimization Algorithm: A Case Study. *Energy Rep.* **2021**, *7*, 1915–1929. [CrossRef]
64. Montana, F.; Kanafani, K.; Wittchen, K.; Birgisdottir, H.; Longo, S.; Cellura, M.; Riva Sanseverino, E. Multi-Objective Optimization of Building Life Cycle Performance. A Housing Renovation Case Study in Northern Europe. *Sustainability* **2020**, *12*, 7807. [CrossRef]
65. Rosso, F.; Ciancio, V.; Dell'Olmo, J.; Salata, F. Multi-Objective Optimization of Building Retrofit in the Mediterranean Climate by Means of Genetic Algorithm Application. *Energy Build.* **2020**, *216*, 109945. [CrossRef]
66. Kadrić, D.; Aganović, A.; Kadrić, E. Multi-Objective Optimization of Energy-Efficient Retrofitting Strategies for Single-Family Residential Homes: Minimizing Energy Consumption, CO_2 Emissions and Retrofit Costs. *Energy Rep.* **2023**, *10*, 1968–1981. [CrossRef]
67. Gao, B.; Zhu, X.; Ren, J.; Ran, J.; Kim, M.K.; Liu, J. Multi-Objective Optimization of Energy-Saving Measures and Operation Parameters for a Newly Retrofitted Building in Future Climate Conditions: A Case Study of an Office Building in Chengdu. *Energy Rep.* **2023**, *9*, 2269–2285. [CrossRef]
68. Wu, R.; Mavromatidis, G.; Orehounig, K.; Carmeliet, J. Multiobjective Optimisation of Energy Systems and Building Envelope Retrofit in a Residential Community. *Appl. Energy* **2017**, *190*, 634–649. [CrossRef]

69. Schütz, T.; Schiffer, L.; Harb, H.; Fuchs, M.; Müller, D. Optimal Design of Energy Conversion Units and Envelopes for Residential Building Retrofits Using a Comprehensive MILP Model. *Appl. Energy* **2017**, *185*, 1–15. [CrossRef]
70. Huang, H.; Wang, H.; Hu, Y.-J.; Li, C.; Wang, X. Optimal Plan for Energy Conservation and CO2 Emissions Reduction of Public Buildings Considering Users' Behavior: Case of China. *Energy* **2022**, *261*, 125037. [CrossRef]
71. Panagiotidou, M.; Aye, L.; Rismanchi, B. Optimisation of Multi-Residential Building Retrofit, Cost-Optimal and Net-Zero Emission Targets. *Energy Build.* **2021**, *252*, 111385. [CrossRef]
72. Dou, Z.; Jin, L.; Chen, Y.; Huang, Z. Optimization of Cost–Carbon Reduction–Technology Solution for Existing Office Parks Based on Genetic Algorithm. *Processes* **2023**, *11*, 2452. [CrossRef]
73. Xu, Y.; Yan, C.; Wang, G.; Shi, J.; Sheng, K.; Li, J.; Jiang, Y. Optimization Research on Energy-Saving and Life-Cycle Decarbonization Retrofitting of Existing School Buildings: A Case Study of a School in Nanjing. *Sol. Energy* **2023**, *254*, 54–66. [CrossRef]
74. Karatas, A.; El-Rayes, K. Optimizing Tradeoffs among Housing Sustainability Objectives. *Autom. Constr.* **2015**, *53*, 83–94. [CrossRef]
75. Schwartz, Y.; Raslan, R.; Mumovic, D. Refurbish or Replace? The Life Cycle Carbon Footprint and Life Cycle Cost of Refurbished and New Residential Archetype Buildings in London. *Energy* **2022**, *248*, 123585. [CrossRef]
76. Sharif, S.A.; Hammad, A. Simulation-Based Multi-Objective Optimization of Institutional Building Renovation Considering Energy Consumption, Life-Cycle Cost and Life-Cycle Assessment. *J. Build. Eng.* **2019**, *21*, 429–445. [CrossRef]
77. Saad, M.M.; Menon, R.P.; Eicker, U. Supporting Decision Making for Building Decarbonization: Developing Surrogate Models for Multi-Criteria Building Retrofitting Analysis. *Energies* **2023**, *16*, 6030. [CrossRef]
78. Wang, P.; Gong, G.; Wang, Y.; Li, L. Thermodynamic Investigation of Building Integrated Energy Efficiency for Building Retrofit. *Energy Build.* **2014**, *77*, 139–148. [CrossRef]
79. Galimshina, A.; Moustapha, M.; Hollberg, A.; Padey, P.; Lasvaux, S.; Sudret, B.; Habert, G. What Is the Optimal Robust Environmental and Cost-Effective Solution for Building Renovation? Not the Usual One. *Energy Build.* **2021**, *251*, 111329. [CrossRef]
80. Mei, J.; Xu, G.; Ahmad, W.; Khan, K.; Amin, M.N.; Aslam, F.; Alaskar, A. Promoting Sustainable Materials Using Recycled Rubber in Concrete: A Review. *J. Clean. Prod.* **2022**, *373*, 133927. [CrossRef]
81. De Boeck, L.; Verbeke, S.; Audenaert, A.; De Mesmaeker, L. Improving the Energy Performance of Residential Buildings: A Literature Review. *Renew. Sustain. Energy Rev.* **2015**, *52*, 960–975. [CrossRef]
82. Shaamala, A.; Yigitcanlar, T.; Nili, A.; Nyandega, D. Algorithmic Green Infrastructure Optimisation: Review of Artificial Intelligence Driven Approaches for Tackling Climate Change. *Sustain. Cities Soc.* **2024**, *101*, 105182. [CrossRef]
83. Ellegaard, O.; Wallin, J.A. The Bibliometric Analysis of Scholarly Production: How Great Is the Impact? *Scientometrics* **2015**, *105*, 1809–1831. [CrossRef] [PubMed]
84. Cobo, M.J.; López-Herrera, A.G.; Herrera-Viedma, E.; Herrera, F. Science Mapping Software Tools: Review, Analysis, and Cooperative Study among Tools. *J. Am. Soc. Inf. Sci. Technol.* **2011**, *62*, 1382–1402. [CrossRef]
85. Chen, C. CiteSpace II: Detecting and Visualizing Emerging Trends and Transient Patterns in Scientific Literature. *J. Am. Soc. Inf. Sci. Technol.* **2006**, *57*, 359–377. [CrossRef]
86. Tetteh, M.O.; Darko, A.; Chan, A.P.C.; Jafari, A.; Brilakis, I.; Chen, W.; Nani, G.; Kwame Yevu, S. Scientometric Mapping of Global Research on Green Retrofitting of Existing Buildings (GREB): Pathway towards a Holistic GREB Framework. *Energy Build.* **2022**, *277*, 112532. [CrossRef]
87. Zhang, N.; Yang, C.; Wang, S. Research Progress and Prospect of Environmental, Social and Governance: A Systematic Literature Review and Bibliometric Analysis. *J. Clean. Prod.* **2024**, *447*, 141489. [CrossRef]
88. Debrah, C.; Chan, A.P.C.; Darko, A. Artificial Intelligence in Green Building. *Autom. Constr.* **2022**, *137*, 104192. [CrossRef]
89. Zhao, H.; Magoulès, F. A Review on the Prediction of Building Energy Consumption. *Renew. Sustain. Energy Rev.* **2012**, *16*, 3586–3592. [CrossRef]
90. Shen, J.; Krietemeyer, B.; Bartosh, A.; Gao, Z.; Zhang, J. Green Design Studio: A Modular-Based Approach for High-Performance Building Design. *Build. Simul.* **2021**, *14*, 241–268. [CrossRef]
91. DesignBuilder Software Ltd-Home. Available online: https://designbuilder.co.uk/ (accessed on 21 February 2024).
92. OpenStudio. Available online: https://openstudio.net/ (accessed on 21 February 2024).
93. 3D Design Software | 3D Modeling on the Web. Available online: http://www.sketchup.com/page/homepage (accessed on 21 February 2024).
94. Ladybug Tools | Honeybee. Available online: https://www.ladybug.tools/honeybee.html (accessed on 11 March 2024).
95. Network, S.D. Created This N. Grasshopper. Available online: https://www.grasshopper3d.com/ (accessed on 21 February 2024).
96. Li, C.Z.; Zhang, L.; Liang, X.; Xiao, B.; Tam, V.W.Y.; Lai, X.; Chen, Z. Advances in the Research of Building Energy Saving. *Energy Build.* **2022**, *254*, 111556. [CrossRef]
97. Pan, Y.; Zhu, M.; Lv, Y.; Yang, Y.; Liang, Y.; Yin, R.; Yang, Y.; Jia, X.; Wang, X.; Zeng, F.; et al. Building Energy Simulation and Its Application for Building Performance Optimization: A Review of Methods, Tools, and Case Studies. *Adv. Appl. Energy* **2023**, *10*, 100135. [CrossRef]
98. Guo, J.; Liu, R.; Xia, T.; Pouramini, S. Energy Model Calibration in an Office Building by an Optimization-Based Method. *Energy Rep.* **2021**, *7*, 4397–4411. [CrossRef]

99. ASHRAE. *Guideline 14, Measurement of Energy and Demand Savings*; American Society of Heating, Refrigerating, and Air Conditioning Engineers: Atlanta, GA, USA, 2002.
100. EVO. *International Performance Measurement and Verification Protocol: Concepts and Options for Determining Energy and Water Savings*; Efficiency Valuation Organization: Washington, DC, USA, 2002; Volume 1.
101. US DOE FEMP. *M&V Guidelines: Measurement and Verification for Performance-Based Contracts, Version 4.0*; Energy Efficiency and Renewable Energy; US DOE FEMP: Washington, DC, USA, 2015.
102. Hong, T.; Kim, J.; Koo, C. LCC and $LCCO_2$ Analysis of Green Roofs in Elementary Schools with Energy Saving Measures. *Energy Build.* **2012**, *45*, 229–239. [CrossRef]
103. Coakley, D.; Raftery, P.; Keane, M. A Review of Methods to Match Building Energy Simulation Models to Measured Data. *Renew. Sustain. Energy Rev.* **2014**, *37*, 123–141. [CrossRef]
104. Nguyen, A.-T.; Reiter, S.; Rigo, P. A Review on Simulation-Based Optimization Methods Applied to Building Performance Analysis. *Appl. Energy* **2014**, *113*, 1043–1058. [CrossRef]
105. Tian, W. A Review of Sensitivity Analysis Methods in Building Energy Analysis. *Renew. Sustain. Energy Rev.* **2013**, *20*, 411–419. [CrossRef]
106. Hou, D.; Hassan, I.G.; Wang, L. Review on Building Energy Model Calibration by Bayesian Inference. *Renew. Sustain. Energy Rev.* **2021**, *143*, 110930. [CrossRef]
107. Huang, H.; Wang, H.; Hu, Y.-J.; Li, C.; Wang, X. The Development Trends of Existing Building Energy Conservation and Emission Reduction—A Comprehensive Review. *Energy Rep.* **2022**, *8*, 13170–13188. [CrossRef]
108. Torabi Moghadam, S.; Toniolo, J.; Mutani, G.; Lombardi, P. A GIS-Statistical Approach for Assessing Built Environment Energy Use at Urban Scale. *Sustain. Cities Soc.* **2018**, *37*, 70–84. [CrossRef]
109. Yussuf, R.O.; Asfour, O.S. Applications of Artificial Intelligence for Energy Efficiency throughout the Building Lifecycle: An Overview. *Energy Build.* **2024**, *305*, 113903. [CrossRef]
110. Seyedzadeh, S.; Rahimian, F.P.; Oliver, S.; Glesk, I.; Kumar, B. Data Driven Model Improved by Multi-Objective Optimisation for Prediction of Building Energy Loads. *Autom. Constr.* **2020**, *116*, 103188. [CrossRef]
111. Grillone, B.; Danov, S.; Sumper, A.; Cipriano, J.; Mor, G. A Review of Deterministic and Data-Driven Methods to Quantify Energy Efficiency Savings and to Predict Retrofitting Scenarios in Buildings. *Renew. Sustain. Energy Rev.* **2020**, *131*, 110027. [CrossRef]
112. Seyedzadeh, S.; Pour Rahimian, F.; Rastogi, P.; Glesk, I. Tuning Machine Learning Models for Prediction of Building Energy Loads. *Sustain. Cities Soc.* **2019**, *47*, 101484. [CrossRef]
113. Magnier, L.; Haghighat, F. Multiobjective Optimization of Building Design Using TRNSYS Simulations, Genetic Algorithm, and Artificial Neural Network. *Build. Environ.* **2010**, *45*, 739–746. [CrossRef]
114. Alwosheel, A.; Van Cranenburgh, S.; Chorus, C.G. Is Your Dataset Big Enough? Sample Size Requirements When Using Artificial Neural Networks for Discrete Choice Analysis. *J. Choice Model.* **2018**, *28*, 167–182. [CrossRef]
115. Liu, C.; Sun, B.; Zhang, C.; Li, F. A Hybrid Prediction Model for Residential Electricity Consumption Using Holt-Winters and Extreme Learning Machine. *Appl. Energy* **2020**, *275*, 115383. [CrossRef]
116. Van Gelder, L.; Das, P.; Janssen, H.; Roels, S. Comparative Study of Metamodelling Techniques in Building Energy Simulation: Guidelines for Practitioners. *Simul. Model. Pract. Theory* **2014**, *49*, 245–257. [CrossRef]
117. Claesen, M.; De Moor, B. Hyperparameter Search in Machine Learning. *arXiv* **2015**, arXiv:1502.02127.
118. Rätz, M.; Javadi, A.P.; Baranski, M.; Finkbeiner, K.; Müller, D. Automated Data-Driven Modeling of Building Energy Systems via Machine Learning Algorithms. *Energy Build.* **2019**, *202*, 109384. [CrossRef]
119. Chen, Z.; Chen, Y.; Xiao, T.; Wang, H.; Hou, P. A Novel Short-Term Load Forecasting Framework Based on Time-Series Clustering and Early Classification Algorithm. *Energy Build.* **2021**, *251*, 111375. [CrossRef]
120. Amasyali, K.; El-Gohary, N.M. A Review of Data-Driven Building Energy Consumption Prediction Studies. *Renew. Sustain. Energy Rev.* **2018**, *81*, 1192–1205. [CrossRef]
121. Al-Shargabi, A.A.; Almhafdy, A.; Ibrahim, D.M.; Alghieth, M.; Chiclana, F. Buildings' Energy Consumption Prediction Models Based on Buildings' Characteristics: Research Trends, Taxonomy, and Performance Measures. *J. Build. Eng.* **2022**, *54*, 104577. [CrossRef]
122. Abbasabadi, N.; Ashayeri, M.; Azari, R.; Stephens, B.; Heidarinejad, M. An Integrated Data-Driven Framework for Urban Energy Use Modeling (UEUM). *Appl. Energy* **2019**, *253*, 113550. [CrossRef]
123. Chen, Y.; Guo, M.; Chen, Z.; Chen, Z.; Ji, Y. Physical Energy and Data-Driven Models in Building Energy Prediction: A Review. *Energy Rep.* **2022**, *8*, 2656–2671. [CrossRef]
124. Venkatraj, V.; Dixit, M.K. Challenges in Implementing Data-Driven Approaches for Building Life Cycle Energy Assessment: A Review. *Renew. Sustain. Energy Rev.* **2022**, *160*, 112327. [CrossRef]
125. Fenner, A.E.; Kibert, C.J.; Li, J.; Razkenari, M.A.; Hakim, H.; Lu, X.; Kouhirostami, M.; Sam, M. Embodied, Operation, and Commuting Emissions: A Case Study Comparing the Carbon Hotspots of an Educational Building. *J. Clean. Prod.* **2020**, *268*, 122081. [CrossRef]
126. Chen, T.Y.; Burnett, J.; Chau, C.K. Analysis of Embodied Energy Use in the Residential Building of Hong Kong. *Energy* **2001**, *26*, 323–340. [CrossRef]
127. Elaouzy, Y.; El Fadar, A. Energy, Economic and Environmental Benefits of Integrating Passive Design Strategies into Buildings: A Review. *Renew. Sustain. Energy Rev.* **2022**, *167*, 112828. [CrossRef]

128. Lin, M.; Ali, A.; Andargie, M.S.; Azar, E. Multidomain Drivers of Occupant Comfort, Productivity, and Well-Being in Buildings: Insights from an Exploratory and Explanatory Analysis. *J. Manag. Eng.* **2021**, *37*, 04021020. [CrossRef]
129. He, C.; Hou, Y.; Ding, L.; Li, P. Visualized Literature Review on Sustainable Building Renovation. *J. Build. Eng.* **2021**, *44*, 102622. [CrossRef]
130. Evins, R. A Review of Computational Optimisation Methods Applied to Sustainable Building Design. *Renew. Sustain. Energy Rev.* **2013**, *22*, 230–245. [CrossRef]
131. Hashempour, N.; Taherkhani, R.; Mahdikhani, M. Energy Performance Optimization of Existing Buildings: A Literature Review. *Sustain. Cities Soc.* **2020**, *54*, 101967.
132. Urbikain, M.K. Energy Efficient Solutions for Retrofitting a Residential Multi-Storey Building with Vacuum Insulation Panels and Low-E Windows in Two European Climates. *J. Clean. Prod.* **2020**, *269*, 121459. [CrossRef]
133. Liu, C.; Sharples, S.; Mohammadpourkarbasi, H. A Review of Building Energy Retrofit Measures, Passive Design Strategies and Building Regulation for the Low Carbon Development of Existing Dwellings in the Hot Summer–Cold Winter Region of China. *Energies* **2023**, *16*, 4115. [CrossRef]
134. Hamby, D.M. A Review of Techniques for Parameter Sensitivity Analysis of Environmental Models. *Environ. Monit. Assess.* **1994**, *32*, 135–154. [CrossRef]
135. Kheiri, F. A Review on Optimization Methods Applied in Energy-Efficient Building Geometry and Envelope Design. *Renew. Sustain. Energy Rev.* **2018**, *92*, 897–920. [CrossRef]
136. Heiselberg, P.; Brohus, H.; Hesselholt, A.; Rasmussen, H.; Seinre, E.; Thomas, S. Application of Sensitivity Analysis in Design of Sustainable Buildings. *Renew. Energy* **2009**, *34*, 2030–2036. [CrossRef]
137. Tian, Z.; Zhang, X.; Jin, X.; Zhou, X.; Si, B.; Shi, X. Towards Adoption of Building Energy Simulation and Optimization for Passive Building Design: A Survey and a Review. *Energy Build.* **2018**, *158*, 1306–1316. [CrossRef]
138. Asadi, E.; Silva, M.G.D.; Antunes, C.H.; Dias, L.; Glicksman, L. Multi-Objective Optimization for Building Retrofit: A Model Using Genetic Algorithm and Artificial Neural Network and an Application. *Energy Build.* **2014**, *81*, 444–456. [CrossRef]
139. Tian, W.; Heo, Y.; De Wilde, P.; Li, Z.; Yan, D.; Park, C.S.; Feng, X.; Augenbroe, G. A Review of Uncertainty Analysis in Building Energy Assessment. *Renew. Sustain. Energy Rev.* **2018**, *93*, 285–301. [CrossRef]
140. Kim, Y.-J.; Ahn, K.-U.; Park, C.-S. Decision Making of HVAC System Using Bayesian Markov Chain Monte Carlo Method. *Energy Build.* **2014**, *72*, 112–121. [CrossRef]
141. Longo, S.; Montana, F.; Riva Sanseverino, E. A Review on Optimization and Cost-Optimal Methodologies in Low-Energy Buildings Design and Environmental Considerations. *Sustain. Cities Soc.* **2019**, *45*, 87–104. [CrossRef]
142. Huang, Y.; Niu, J. Optimal Building Envelope Design Based on Simulated Performance: History, Current Status and New Potentials. *Energy Build.* **2016**, *117*, 387–398. [CrossRef]
143. Deb, K.; Pratap, A.; Agarwal, S.; Meyarivan, T. A Fast and Elitist Multiobjective Genetic Algorithm: NSGA-II. *IEEE Trans. Evol. Comput.* **2002**, *6*, 182–197. [CrossRef]
144. Liu, Y.; Heuvelink, G.B.M.; Bai, Z.; He, P.; Xu, X.; Ding, W.; Huang, S. Analysis of Spatio-Temporal Variation of Crop Yield in China Using Stepwise Multiple Linear Regression. *Field Crops Res.* **2021**, *264*, 108098. [CrossRef]
145. Ali, A.; Jayaraman, R.; Azar, E.; Maalouf, M. A Comparative Analysis of Machine Learning and Statistical Methods for Evaluating Building Performance: A Systematic Review and Future Benchmarking Framework. *Build. Environ.* **2024**, *252*, 111268. [CrossRef]
146. Amini Toosi, H.; Lavagna, M.; Leonforte, F.; Del Pero, C.; Aste, N. Life Cycle Sustainability Assessment in Building Energy Retrofitting; A Review. *Sustain. Cities Soc.* **2020**, *60*, 102248. [CrossRef]
147. Mousavi, S.; Villarreal-Marroquín, M.G.; Hajiaghaei-Keshteli, M.; Smith, N.R. Data-Driven Prediction and Optimization toward Net-Zero and Positive-Energy Buildings: A Systematic Review. *Build. Environ.* **2023**, *242*, 110578. [CrossRef]
148. Machairas, V.; Tsangrassoulis, A.; Axarli, K. Algorithms for Optimization of Building Design: A Review. *Renew. Sustain. Energy Rev.* **2014**, *31*, 101–112. [CrossRef]
149. Congradac, V.; Kulic, F. HVAC System Optimization with CO_2 Concentration Control Using Genetic Algorithms. *Energy Build.* **2009**, *41*, 571–577. [CrossRef]
150. Delgarm, N.; Sajadi, B.; Delgarm, S.; Kowsary, F. A Novel Approach for the Simulation-Based Optimization of the Buildings Energy Consumption Using NSGA-II: Case Study in Iran. *Energy Build.* **2016**, *127*, 552–560. [CrossRef]
151. Deb, K.; Thiele, L.; Laumanns, M.; Zitzler, E. Scalable Test Problems for Evolutionary Multiobjective Optimization. In *Evolutionary Multiobjective Optimization*; Abraham, A., Jain, L., Goldberg, R., Eds.; Advanced Information and Knowledge Processing; Springer: London, UK, 2005; pp. 105–145. ISBN 978-1-85233-787-2.
152. Deb, K.; Jain, H. An Evolutionary Many-Objective Optimization Algorithm Using Reference-Point-Based Nondominated Sorting Approach, Part I: Solving Problems with Box Constraints. *IEEE Trans. Evol. Comput.* **2014**, *18*, 577–601. [CrossRef]
153. Emmerich, M.T.M.; Deutz, A.H. A Tutorial on Multiobjective Optimization: Fundamentals and Evolutionary Methods. *Nat. Comput.* **2018**, *17*, 585–609. [CrossRef] [PubMed]
154. Mendoza, F.; Bernal-Agustin, J.L.; Dominguez-Navarro, J.A. NSGA and SPEA Applied to Multiobjective Design of Power Distribution Systems. *IEEE Trans. Power Syst.* **2006**, *21*, 1938–1945. [CrossRef]
155. Zitzler, E.; Laumanns, M.; Thiele, L. *SPEA2: Improving the Strength Pareto Evolutionary Algorithm*; ETH Zurich: Zürich, Switzerland, 2001; p. 21.

156. Liu, S.; Tao, R.; Tam, C.M. Optimizing Cost and CO_2 Emission for Construction Projects Using Particle Swarm Optimization. *Habitat Int.* **2013**, *37*, 155–162. [CrossRef]
157. Gan, V.J.L.; Lo, I.M.C.; Ma, J.; Tse, K.T.; Cheng, J.C.P.; Chan, C.M. Simulation Optimisation towards Energy Efficient Green Buildings: Current Status and Future Trends. *J. Clean. Prod.* **2020**, *254*, 120012. [CrossRef]
158. Raschka, S.; Patterson, J.; Nolet, C. Machine Learning in Python: Main Developments and Technology Trends in Data Science, Machine Learning, and Artificial Intelligence. *Information* **2020**, *11*, 193. [CrossRef]
159. MATLAB. Available online: https://www.mathworks.com/products/matlab.html (accessed on 7 April 2024).
160. Start [Jeplus.Org]. Available online: http://www.jeplus.org/wiki/doku.php?id=start (accessed on 7 April 2024).
161. Multiobjective Optimization-MATLAB & Simulink. Available online: https://www.mathworks.com/help/gads/multiobjective-optimization.html (accessed on 7 April 2024).
162. Cui, Y.; Geng, Z.; Zhu, Q.; Han, Y. Review: Multi-Objective Optimization Methods and Application in Energy Saving. *Energy* **2017**, *125*, 681–704. [CrossRef]
163. Veldhuizen, D.A.V.; Lamont, G.B. *Multiobjective Evolutionary Algorithm Research: A History and Analysis*; TR-98e03; Air Force Institute of Technology: Dayton, OH, USA, 1998.
164. Zitzler, E.; Thiele, L.; Laumanns, M.; Fonseca, C.M.; Da Fonseca, V.G. Performance Assessment of Multiobjective Optimizers: An Analysis and Review. *IEEE Trans. Evol. Comput.* **2003**, *7*, 117–132. [CrossRef]
165. Pinzon Amorocho, J.A.; Hartmann, T. A Multi-Criteria Decision-Making Framework for Residential Building Renovation Using Pairwise Comparison and TOPSIS Methods. *J. Build. Eng.* **2022**, *53*, 104596. [CrossRef]
166. Chauhan, A.; Vaish, R. A Comparative Study on Decision Making Methods with Interval Data. *J. Comput. Eng.* **2014**, *2014*, 93074. [CrossRef]
167. Rokbani, N.; Neji, B.; Slim, M.; Mirjalili, S.; Ghandour, R. A Multi-Objective Modified PSO for Inverse Kinematics of a 5-DOF Robotic Arm. *Appl. Sci.* **2022**, *12*, 7091. [CrossRef]
168. Francart, N.; Malmqvist, T.; Hagbert, P. Climate Target Fulfilment in Scenarios for a Sustainable Swedish Built Environment beyond Growth. *Futures* **2018**, *98*, 1–18. [CrossRef]
169. Ahmad, M.W.; Mourshed, M.; Rezgui, Y. Trees vs Neurons: Comparison between Random Forest and ANN for High-Resolution Prediction of Building Energy Consumption. *Energy Build.* **2017**, *147*, 77–89. [CrossRef]
170. Wang, Z.; Liu, J.; Zhang, Y.; Yuan, H.; Zhang, R.; Srinivasan, R.S. Practical Issues in Implementing Machine-Learning Models for Building Energy Efficiency: Moving beyond Obstacles. *Renew. Sustain. Energy Rev.* **2021**, *143*, 110929. [CrossRef]
171. Singaravel, S.; Suykens, J.; Geyer, P. Deep-Learning Neural-Network Architectures and Methods: Using Component-Based Models in Building-Design Energy Prediction. *Adv. Eng. Inform.* **2018**, *38*, 81–90. [CrossRef]
172. Khalil, M.; McGough, A.S.; Pourmirza, Z.; Pazhoohesh, M.; Walker, S. Machine Learning, Deep Learning and Statistical Analysis for Forecasting Building Energy Consumption—A Systematic Review. *Eng. Appl. Artif. Intell.* **2022**, *115*, 105287. [CrossRef]
173. Qian, F.; Gao, W.; Yang, Y.; Yu, D. Potential Analysis of the Transfer Learning Model in Short and Medium-Term Forecasting of Building HVAC Energy Consumption. *Energy* **2020**, *193*, 116724. [CrossRef]
174. Lu, C.; Li, S.; Lu, Z. Building Energy Prediction Using Artificial Neural Networks: A Literature Survey. *Energy Build.* **2022**, *262*, 111718. [CrossRef]
175. Man, Q.; Yu, H.; Feng, K.; Olofsson, T.; Lu, W. Transfer of Building Retrofitting Evaluations for Data-Scarce Conditions: An Empirical Study for Sweden to China. *Energy Build.* **2024**, *310*, 114041. [CrossRef]
176. Fahlstedt, O.; Temeljotov-Salaj, A.; Lohne, J.; Bohne, R.A. Holistic Assessment of Carbon Abatement Strategies in Building Refurbishment Literature—A Scoping Review. *Renew. Sustain. Energy Rev.* **2022**, *167*, 112636. [CrossRef]
177. Bian, Y.; Chen, Y.; Sun, Y.; Ma, Y.; Yu, D.; Leng, T. Simulation of Daylight Availability, Visual Comfort and View Clarity for a Novel Window System with Switchable Blinds in Classrooms. *Build. Environ.* **2023**, *235*, 110243. [CrossRef]
178. Xu, S.; Li, C.; He, W.; Chu, W.; Hu, Z.; Lu, B. Experimental Study of Bifacial Photovoltaic Wall System Incorporating Thermochromic Material. *Sustain. Cities Soc.* **2024**, *106*, 105372. [CrossRef]
179. Bui, D.-K.; Nguyen, T.N.; Ghazlan, A.; Ngo, T.D. Biomimetic Adaptive Electrochromic Windows for Enhancing Building Energy Efficiency. *Appl. Energy* **2021**, *300*, 117341. [CrossRef]
180. Cheng, Y.; Chen, X.; Xu, B.; Pei, G.; Jiao, D. Experimental Analysis of Building Envelope Integrating Phase Change Material and Cool Paint under a Real Environment in Autumn. *J. Clean. Prod.* **2024**, *461*, 142674. [CrossRef]
181. Quan, S.J.; Wu, J.; Wang, Y.; Shi, Z.; Yang, T.; Yang, P.P.-J. Urban Form and Building Energy Performance in Shanghai Neighborhoods. *Energy Procedia* **2016**, *88*, 126–132. [CrossRef]
182. Ekici, B.; Cubukcuoglu, C.; Turrin, M.; Sariyildiz, I.S. Performative Computational Architecture Using Swarm and Evolutionary Optimisation: A Review. *Build. Environ.* **2019**, *147*, 356–371. [CrossRef]
183. Octopus. Available online: https://www.grasshopper3d.com/groups/group/show?groupUrl=octopus (accessed on 7 April 2024).
184. Fan, Z.; Liu, M.; Tang, S. A Multi-Objective Optimization Design Method for Gymnasium Facade Shading Ratio Integrating Energy Load and Daylight Comfort. *Build. Environ.* **2022**, *207*, 108527. [CrossRef]
185. Nasrollahzadeh, N. Comprehensive Building Envelope Optimization: Improving Energy, Daylight, and Thermal Comfort Performance of the Dwelling Unit. *J. Build. Eng.* **2021**, *44*, 103418. [CrossRef]
186. Ji, Y.; Wang, W.; He, Y.; Li, L.; Zhang, H.; Zhang, T. Performance in Generation: An Automatic Generalizable Generative-Design-Based Performance Optimization Framework for Sustainable Building Design. *Energy Build.* **2023**, *298*, 113512. [CrossRef]

187. Chen, R.; Samuelson, H.; Zou, Y.; Zheng, X.; Cao, Y. Improving Building Resilience in the Face of Future Climate Uncertainty: A Comprehensive Framework for Enhancing Building Life Cycle Performance. *Energy Build.* **2024**, *302*, 113761. [CrossRef]
188. Madushika, U.G.D.; Lu, W. Green Retrofitting Application in Developing Economies: State of the Art and Future Research Directions. *Energy Build.* **2023**, *301*, 113712. [CrossRef]
189. Tian, Z.; Zhang, X.; Wei, S.; Du, S.; Shi, X. A Review of Data-Driven Building Performance Analysis and Design on Big on-Site Building Performance Data. *J. Build. Eng.* **2021**, *41*, 102706. [CrossRef]
190. Singh, M.K.; Attia, S.; Mahapatra, S.; Teller, J. Assessment of Thermal Comfort in Existing Pre-1945 Residential Building Stock. *Energy* **2016**, *98*, 122–134. [CrossRef]
191. Alonso, A.; Patricio, J.; Suárez, R.; Escandón, R. Acoustical Retrofit of Existing Residential Buildings: Requirements and Recommendations for Sound Insulation between Dwellings in Europe and Other Countries Worldwide. *Build. Environ.* **2020**, *174*, 106771. [CrossRef]
192. Aftab, M.; Chen, C.; Chau, C.-K.; Rahwan, T. Automatic HVAC Control with Real-Time Occupancy Recognition and Simulation-Guided Model Predictive Control in Low-Cost Embedded System. *Energy Build.* **2017**, *154*, 141–156. [CrossRef]
193. Hong, Y.; Ezeh, C.I.; Deng, W.; Hong, S.-H.; Peng, Z. Building Energy Retrofit Measures in Hot-Summer–Cold-Winter Climates: A Case Study in Shanghai. *Energies* **2019**, *12*, 3393. [CrossRef]
194. Li, L.; Ma, Q.; Gao, W.; Wei, X. Incorporating Users' Adaptive Behaviors into Multi-Objective Optimization of Shading Devices: A Case Study of an Office Room in Qingdao. *Energy Build.* **2023**, *301*, 113683. [CrossRef]
195. De Wilde, P.; Coley, D. The Implications of a Changing Climate for Buildings. *Build. Environ.* **2012**, *55*, 1–7. [CrossRef]
196. Daly, D.; Cooper, P.; Ma, Z. Implications of Global Warming for Commercial Building Retrofitting in Australian Cities. *Build. Environ.* **2014**, *74*, 86–95. [CrossRef]
197. Balezentis, T.; Siksnelyte-Butkiene, I.; Streimikiene, D. Stakeholder Involvement for Sustainable Energy Development Based on Uncertain Group Decision Making: Prioritizing the Renewable Energy Heating Technologies and the BWM-WASPAS-IN Approach. *Sustain. Cities Soc.* **2021**, *73*, 103114. [CrossRef]
198. Beccali, M.; Cellura, M.; Fontana, M.; Longo, S.; Mistretta, M. Energy Retrofit of a Single-Family House: Life Cycle Net Energy Saving and Environmental Benefits. *Renew. Sustain. Energy Rev.* **2013**, *27*, 283–293. [CrossRef]
199. Allegrini, J.; Dorer, V.; Carmeliet, J. Influence of the Urban Microclimate in Street Canyons on the Energy Demand for Space Cooling and Heating of Buildings. *Energy Build.* **2012**, *55*, 823–832. [CrossRef]
200. Santamouris, M.; Papanikolaou, N.; Livada, I.; Koronakis, I.; Georgakis, C.; Argiriou, A.; Assimakopoulos, D.N. On the Impact of Urban Climate on the Energy Consumption of Buildings. *Sol. Energy* **2001**, *70*, 201–216. [CrossRef]
201. Reitberger, R.; Palm, N.; Palm, H.; Lang, W. Urban Systems Exploration: A Generic Process for Multi-Objective Urban Planning to Support Decision Making in Early Design Phases. *Build. Environ.* **2024**, *254*, 111360. [CrossRef]

energies

MDPI

Review

A Critical Review of Overheating Risk Assessment Criteria in International and National Regulations—Gaps and Suggestions for Improvements

Mahsan Sadeghi [1,2,3,*], Dong Chen [1] and Anthony Wright [1]

1 Commonwealth Scientific Industrial Research Organisation (CSIRO), Melbourne 3168, Australia; dong.chen@csiro.au (D.C.); anthony.wright@csiro.au (A.W.)
2 IEQ Lab, School of Architecture, Design, and Planning, The University of Sydney, Sydney 2006, Australia
3 NHMRC Centre for Safe Air, The University of Tasmania, Hobart 7005, Australia
* Correspondence: mahsan.sadeghi@csiro.au

Abstract: The escalating environmental threat of indoor overheating, exacerbated by global climate change, urbanisation, and population growth, poses a severe risk to public health worldwide, specifically to those regions which are exposed to extreme heat events, such as Australia. This study delves into the critical issue of overheating within residential buildings, examining the existing state of knowledge on overheating criteria and reviewing overheating guidelines embedded in (a) international standards and (b) national building codes. Each regulatory document is analysed based on its underlying thermal comfort model, metric, and indices. The advantages and limitations of each document are practically discussed and for each legislative document and standard, and the quantitative measures have been reviewed, analysed, and summarised. The findings illuminate a global reliance on simplistic indices, such as indoor air temperature and operative temperature, in the existing regulatory documents. However, other critical environmental parameters, such as relative humidity, indoor air velocity, and physiological parameters including metabolic heat production and clothing insulation, are often not included. The absence of mandatory regulations for overheating criteria in residential buildings in some countries, such as in Australian homes, prompts the call for a holistic approach based on a thermal index inclusive of relevant environmental and physiological parameters to quantify heat stress exposure based on human thermal regulation. Gaps and limitations within existing guidelines are identified, and recommendations are proposed to strengthen the regulatory framework for overheating risk assessment in residential buildings. The findings hold significance for policymakers, building energy assessors, architects, and public health professionals, providing direction for the improvement of existing, and development of new, guidelines that aim to enhance indoor thermal condition and population health while ensuring energy efficiency and sustainability in the building stock.

Keywords: overheating; residential buildings; international standards; national building codes; *PMV*; adaptive thermal comfort; building simulation

1. Introduction

Overheating is a critical environmental hazard affecting billions of people globally and presenting a significant threat to public health [1]. Climate change, urbanisation, and population growth drive rising temperatures in many regions, exacerbating existing health disparities and increasing the risk of heat-related illnesses and deaths [2]. The direct and indirect impacts of overheating are complex, affecting various systems within the human body, from cardiovascular and respiratory function to mental health and cognitive performance (e.g., refs. [3–6]). Impact studies indicate that extreme weather events, such as heat waves and flooding, may occur more frequently due to anthropogenic climate change, drawing attention from policymakers and researchers worldwide [7].

In addition to global and regional climate processes, urban heat island (UHI) is another phenomenon that influences microclimates. Expansion of impervious land surface areas reduced green spaces, complex urban morphology, and anthropogenic heat emissions, and all conspire to intensify urban heat which, in turn, significantly exacerbates heat in indoor environments where we spend the majority of our daily lives [8]. The spill over of external heat into indoor spaces exacerbates overheating challenges, particularly in densely populated urban settings (e.g., Refs. [9–13]).

Extreme events can further amplify existing health disparities and socio-economic inequalities, with low-income communities facing greater exposure and susceptibility to heat-related illnesses and deaths [8]. It behooves us to apply a thorough understanding of the mechanisms underlying overheating impacts to the development of effective strategies to mitigate the increasing threat of extreme heat globally.

Despite the significant risks posed by overheating and heatwaves to public health, there is a current lack of clear and consistent guidelines and standards for ensuring thermal comfort and mitigating overheating risks in some countries around the world, i.e., in Australian residential buildings. Although some international standards and national building codes exist, their applicability to other countries may be limited due to the differences in climate conditions. Figure 1 illustrates the regional distribution of regulatory documents globally, predominantly implemented in North America, Europe, and East Asia (including, China, India, and Singapore). In addition to international standards, some countries, including France, Germany, the United Kingdom, Belgium, Sweden, Austria, Italy, Spain, and Greece, successfully integrated regulations for thermal comfort and overheating into their national building codes and standards [14–24].

Different countries and blocs maintain their own distinct standards, codes, and guidelines for thermal comfort and preventing overheating in residential buildings. For example, the American Society of Heating, Refrigerating, and Air-Conditioning Engineers (ASHRAE) developed ASHRAE Standard 55, offering guidelines for thermal comfort in indoor environments, which has been widely implemented in North America [25]. Meanwhile, the European Union's Energy Performance of Buildings Directive establishes minimum energy performance requirements for buildings, while maintaining minimum levels of indoor environmental quality (IEQ) in general, and indoor thermal comfort in particular [15]. In the United Kingdom, building regulations provide guidance on maximum allowable temperatures in dwellings during hot weather, along with requirements for adequate ventilation and shading [16].

It is also essential to consider the role of building simulation tools in the assessment of thermal comfort conditions. These tools enable detailed analysis of indoor environmental conditions by modelling heat transfer, ventilation, mean radiant temperature (MRT), and occupation behaviour under various scenarios. However, variations in simulation algorithms and assumptions can lead to discrepancies in results, highlighting the need for standardised methodologies to improve consistency and reliability across different tools [26]. It is also important to note that correct prediction of MRT by building simulation tools plays a significant role in predicting human thermal comfort and thermal regulation, which has been overlooked in the literature [26].

Australia is a vast country with variable climate which is classified into eight zones by the Australian Building Codes Board [27]. Despite Australia's unique climate, there is currently a lack of specific standards addressing overheating and thermal comfort in buildings. In the absence of a national overheating standard, analysis in Australia often relies on standards from other countries as indicated by [22]. Therefore, to develop a national overheating criterion, it will be necessary to consider the environmental conditions for each climate zone separately to be able to develop a nuanced, bio-climatically relevant approach to overheating. In this article, our objectives are to (a) review the current state of knowledge on overheating and thermal comfort criteria in residential buildings, focusing on existing national building codes and international standards; (b) identify gaps and limitations in these guidelines; and (c) provide suggestions for improving the regulatory framework. The findings of this manuscript could be used to guide the development of new overheating regulatory documents for countries that have not employed any such as Australia, and those countries which are more sensitive to the increase in temperature, such as southern European countries, with a prospective vision for enhancing the health and well-being of residents while ensuring energy efficiency in buildings.

Figure 1. Regional distribution of regulatory documents around the world. The regulatory documents have mostly been implemented in North America, Europe, and East Asia. Source for the map: ArcGIS® software 10.7.1 (Esri (Environmental Systems Research Institute, Inc.), in Redlands, CA, USA) [28].

2. Review Methodology

This paper reviews overheating criteria and thermal comfort thresholds in (a) international standards and (b) national building codes. The scope of the manuscript has been narrowed down to the residential regulatory documents globally across all available climate zones. Each regulatory document is analysed based on its underlying thermal comfort model, metric, and indices. The advantages and limitations of each document are discussed in this manuscript.

The procedure employed for reviewing the literature was a thematic literature review approach to systematically review existing international standards and national building codes related to residential overheating. The review focused on identifying recurring themes, criteria, and gaps in how overheating is addressed in building regulations globally. Relevant literature was sourced from academic databases, including ScienceDirect, Web of Science, Scopus, and Google Scholar, using targeted keywords and Boolean operators to ensure a comprehensive search. After screening and selecting pertinent studies and documents, the findings were grouped into key themes, such as overheating thresholds, thermal comfort indices, and metrics. Each regulatory document was analysed to synthesise insights, highlight limitations and advantages, and identify areas for improvement in regulatory approaches.

2.1. Standards and Regulatory Documents

Regulatory documents that have been included in this review were listed as:

a. International standards
   I. ASHRAE 55 (2017), and ASHRAE 55 (2020)—Thermal Environmental Conditions for Human Occupancy [29,30]
   II. CIBSE Guide A (2006), and CIBSE Guide A (2015)—Environmental Design [31]
   III. CIBSE TM52—The limits of Thermal Comfort: Avoiding Overheating in European Buildings [18]
   IV. CIBSE TM59 (2017)—Design Methodology for the Assessment of Building Services Engineers [32]
   V. EN 15251 (2007)—Indoor environmental parameters for design and assessment of energy performance of buildings addressing indoor air quality, thermal environment. Lighting and acoustics [19]
   VI. EN 16798 (2019) [33]
   VII. ISO 7730 (2005)—Ergonomics of the thermal environment [34]
   VIII. Passive House (2015) [35]
b. National building codes
   I. Belgium
   II. France
   III. Germany
   IV. Netherlands
   V. UK

2.2. Quantitative Measures

For each legislative document and standard the following quantitative measures have been reviewed, analysed, and summarised:

a. Thermal comfort model (static thermal comfort, or adaptive thermal comfort model).
b. Temporal domain (i.e., occupied hours; sleeping hours; and all hours).
c. Spatial domain (i.e., living room; bedroom; and kitchen).
d. Indices for quantifying thermal comfort (i.e., PMV; T_{Oi}; and T_{ai}).
e. Metrics for heat assessment (i.e., degree hours; hours; and percentage of hours outside the comfort range).
f. Threshold limit (the maximum or minimum acceptable measure).
g. Building operational category (i.e., naturally, mechanically ventilated building, or mixed mode operation building).

3. Findings and Discussions

This section illuminates the findings of the study, reviews the overheating assessment criteria, identifies the gaps, and critically discusses and examines the limitations of existing standards and building codes.

3.1. Thermal Comfort Parameters and Responses

To review and analyse the criteria of overheating, we need to first identify the definition of thermal comfort, thermal discomfort, and the difference between adaptive and static thermal comfort models.

According to ASHRAE-55 standard (2017) [29], thermal comfort refers to the mental state in which an individual expresses satisfaction with their environments' thermal conditions. There are six universally agreed parameters that influence thermal comfort, from which four are environmental parameters and two are so-called "personal parameters". Figure 2 shows the environmental and personal parameters that cause thermal discomfort.

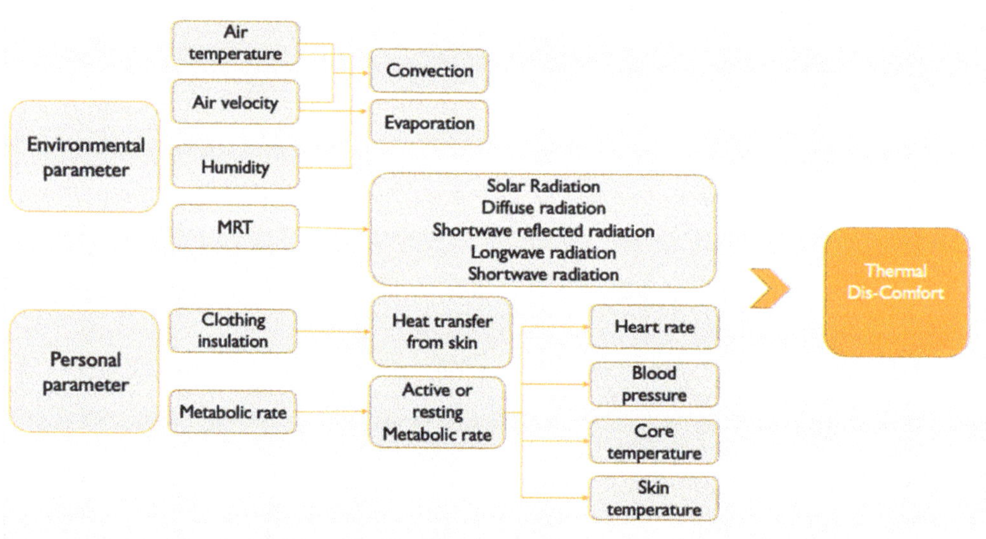

Figure 2. Environmental and personal parameters causing thermal discomfort.

Four environmental parameters, namely air temperature, humidity, air velocity, and mean radiant temperature (MRT), impact human thermal regulation by causing changes in the human body's heat exchanges with its environment via convective, conduction, radiative, and evapourative process. Mean radiant temperature is a key environmental parameter affecting radiative fluxes, exerting a significant impact on human thermal physiology in indoor environments, and even more so in outdoor settings where solar radiation is present. Radiative fluxes refer to the electromagnetic energy emitted by all matter having temperature above absolute zero [36]. It includes shortwave length components of direct and diffuse solar radiation and shortwave reflected radiation. Longwave-length radiation components include atmospheric infrared emission from greenhouse gases (particularly water vapour) and radiation emission from ground and built environmental surfaces [37,38].

Mean radiant temperature refers to the average temperature of all the surfaces surrounding a person, including the walls, sky, floor or ground, and other objects. High mean radiant temperatures due to intense solar radiation in outdoor environments, or hot surfaces in built environments, will lead to overheating and discomfort. Conversely, low mean radiant temperatures will exaggerate radiative heat loss from the body which, if sustained for long enough, lead to reduced skin temperature, shivering, and subjective cold discomfort. Therefore, accurate and effective measurement and management of all these elements in the bioclimatic radiative environment are crucial for ensuring thermal comfort and minimising the risk of thermal health problems, particularly in vulnerable populations such as the elderly and those with pre-existing medical conditions [39].

Human responses to the thermal environment can be categorised into three groups:

(a) Physiological level responses operate under autonomic nervous system control and include vasomotor activity (dilation and constriction), sweating, and shivering [40].

(b) Behavioural level responses are mediated by the central nervous system and include adjustments to the environment such as changing the rate of air movement indoors (Vi) by opening/closing windows, controlling fans, and directly adjusting room temperature with whatever conditioning technology presents. Another class of occupant behavioural response involves directly adjusting to the heat balance personal parameters, such as changing clothing insulation levels or changing the rate of metabolic heat production (e.g., resting after physical exertion) (e.g., refs. [41,42]).

(c) Psychological response refers to comfort expectations which derive from the cultural and climatological background.

Apart from these three categories of human response to the climatic environment is the building itself, including temperature swings in the building fabric. Figure 3 categorises the various types of thermal responses from the building occupant system.

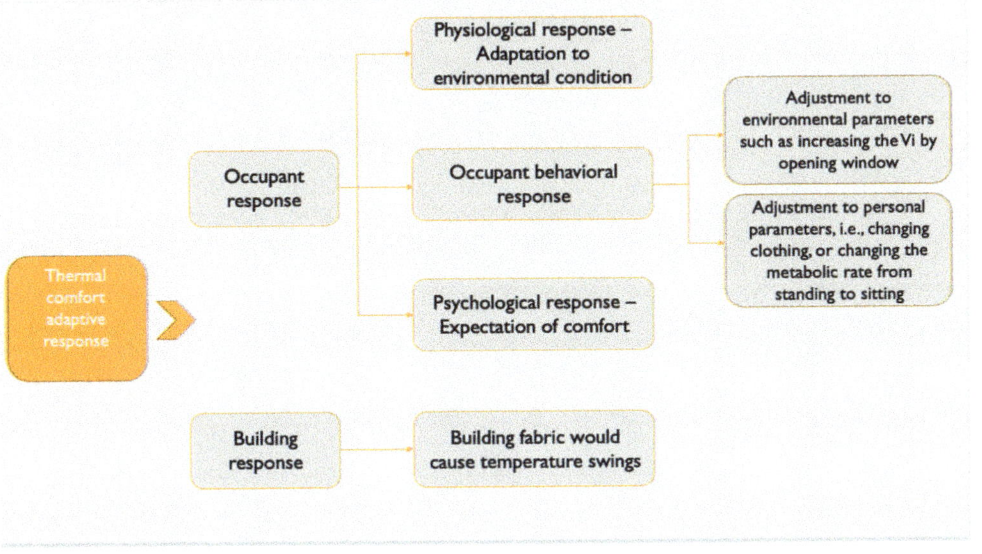

Figure 3. Adaptive response to thermal comfort from occupant and building.

3.2. Static Thermal Comfort Model

Static thermal comfort models underpin the conventional approach to indoor environmental design based on fixed, uniform temperature settings. The aim is to maintain a consistent temperature throughout a space to ensure that the majority of occupants are comfortable. The most widely used indices for predicting static thermal comfort in building design and environmental studies are known as the predicted mean vote (*PMV*) and predicted percentage dissatisfied (*PPD*), based on Fanger's 1970 heat balance "comfort equation" [43].

The *PMV* index expresses thermal comfort of a space in terms of the typical thermal sensation that an "average" person may experience in that space on the basis of the four thermal environmental and two personal thermal parameters enumerated earlier in Section 3.1. *PMV* is widely used to guide design and operational decisions about heating, ventilation, and air conditioning (HVAC) systems, but also building design, façade specifications, and floorplate layout.

The predicted percentage dissatisfied (*PPD*) index is the second part of the Fanger) static comfort model [44] and is itself derived from the *PMV* index. *PMV* refers to the average thermal sensation (expressed on a seven-point scale) of a large group of building occupants exposed to a given combination of indoor climatic and personal thermal parameters. That average is considered to be the mode of a normal (Gaussian) bell-shaped distribution of thermal sensation votes from the entire group along the seven-point sensation scale [45].

The central three categories of that subjective warmth scale (i.e., −1 slightly cool, 0 neutral, and +1 slightly warm) were assumed by Fanger to correspond with acceptable or thermally satisfactory subjective states. *PPD* refers to the area of the Gaussian distribution 'tails' of thermal sensation votes falling beyond the thermally acceptable central three sensation categories (i.e., −3 cold, −2 cool, +2 warm, and +3 hot).

PMV/*PPD* static comfort assumes all occupants have similar preferences and responses to the given thermal conditions. In reality, however, individuals have very diverse sensitivities to temperature and other environmental stimuli. As a result, the *PMV*/*PPD* index may not accurately reflect the thermal comfort of many occupants in a given indoor environment [43]. Most standards recommend the static comfort models for air-conditioned buildings and the adaptive comfort models for non-air-conditioned or NV buildings.

3.3. Adaptive Thermal Comfort

Adaptive thermal comfort is a more recent approach that considers the dynamic nature of indoor environments, outdoor mean temperature, and individual occupant preferences. It recognises that people can tolerate a wider range of temperatures and that thermal comfort is influenced by a variety of adaptive responses and thermal background, such as adjustment to personal and environmental parameters initiated by occupants.

Therefore, the adaptive approach aims to provide a range of thermal conditions and allow occupants to adjust their clothing or behaviour accordingly, resulting in a more comfortable and energy-efficient indoor environment. Adaptive thermal comfort is mostly used for naturally ventilated buildings (NV). Equation (1) shows the concept for calculating adaptive thermal comfort.

$$To_i = (a \times Tarm_{out}) + b \tag{1}$$

where

To_i = Indoor comfort operative temperature (dependent variable);

$Tarm_{out}$ = Outdoor running mean air temperature (independent variable);

a = Gradient, proportional to the degree of adaptation to local climatic condition;

b = Y-intercept.

In ASHRAE-55 standard (2017) [29], a and b are identified as follows (Equation (2)):

$$To_i = 0.31 \times Tarm_{out} + 21.3 \ (\textit{if the air velocity} \leq 0.3 \text{ m/s}). \tag{2}$$

In EN15251 Cat II [17], CIBSE Guide A [31], and ISSO 74 Class C, a and b are identified as Equation (3).

$$To_i = 0.33 \times Tarm_{out} + 21.8 \ (\textit{if the air velocity} \leq 0.3 \text{ m/s}) \tag{3}$$

In Griffiths' Method [46], a and b are described as following (Equation (4)):

$$T_n = 0.35 \times Tarm_{out} + 13.5. \tag{4}$$

where

T_n = Neutral air temperature—indoor.

Modified Griffiths' Method [47,48] describes a and b as Equation (5).

$$T_{accept} = 0.26 \times Tarm_{out} + 15.5 \tag{5}$$

where

T_{accept} = Acceptable temperature.

De dear and Brager [49] developed a revision model for ASHRAE-55 (2020) 31for NV buildings, and illuminated a and b as follows (Equation (6)):

$$T_n = 0.31 \times Tarm_{out} + 17.8. \tag{6}$$

3.4. Analytical and Critical Comparison of International Standards

This section compares the metrics, indices, and threshold limits that have been implemented in the international standards, and critically reviews the advantage and limitations of each document.

3.4.1. EN 15251

Metric

EN 15251 [17] applies three metrics for quantifying the overheating or discomfort. These metrics are:

(a) Percentage of occupied hours outside the range (%*OhOR*).
(b) Degree hours (*Dh*).
(c) Weighted *PPD*.

Percentage of Occupied Hours Outside the Range (%*OhOR*)

The %*OhOR* metric indicates the percentage of hours in which indoor operative temperature (T_{oi}) or the *PMV* falls outside the predefined comfort thresholds associated with the chosen comfort category. This metric is relevant for both static and adaptive comfort models.

To compute the %*OhOR*, a binary weighting factor is employed, where its value is set to zero during periods when the conditions remain within the comfort ranges, and it assumes a value of one when conditions deviate from these ranges (Equation (7)). This index provides a simple and direct representation of the frequency of discomfort, highlighting instances when occupants may experience discomfort due to environmental conditions. EN 15251 suggests acceptable deviation of 3–5% for both short-term and long-term periods, including daily, weekly, monthly, and yearly periods.

$$\%OhOR = \sum_{i=1}^{N\ Toi>Tcom} \tag{7}$$

where

N = number of occupied hours in a day when T_{oi} is above the comfort temperature,
T_{oi} = indoor operative temperature,
T_{com} = comfort temperature.

Degree Hours (*Dh*)

The degree hours (*Dh*) refers to the number of hours which fall above the acceptable threshold, multiplied into number of degrees above the threshold (Equation (8)). The maximum acceptable threshold of *Dh* is not identified for this metric. *Dh* is not normalised to the number of occupied hours.

$$Dh = \sum_{i=1}^{N} Toi - Tcom \tag{8}$$

Weighted *PPD* ($\sum PPD$)

Weighted *PPD* refers to occupied hours, when *PMV* exceeds the comfort threshold (above +0.5, or lower than −0.5). It quantifies the summation of hourly percentage of dissatisfaction (Equation (9)). However, no maximum threshold is identified.

$$PPD = \sum_{i=1}^{N-0.5>PMV>0.5} \tag{9}$$

where

N = number of occupied hours when *PMV* is above or lower comfort threshold.

Index

Indices applied in EN 15251 are indoor operative temperature (T_{oi}) and *PMV*. They could be applied in both adaptive and static thermal comfort models.

The limitations for EN 15251 criteria would be:

(a) The *%OhOR* metric does not consider the intensity of exposure to heat.
(b) The *%OhOR* metric applies the same threshold limitations for short-term (i.e., weekly), and long-term (yearly) exposure to discomfort.
(c) The degree hours metric is not normalised to the occupied hours.
(d) There is not a maximum threshold for the degree hour and weighted *PPD* metrics.
(e) The index operative temperature does not consider air movement and relative humidity.

3.4.2. EN 16798

EN 16798 [18] suggests the same metric and indices used in EN 15251 for the thermal comfort measures; however, it revised the shortcoming of EN 15251 and suggests different acceptable percentage for short term and long-term exposure to heat. It proposes a 20–50% acceptable deviation on a weekly basis, 50% is the maximum acceptable criterion for the weekly basis. Accordingly, it suggests 12–25% for a monthly basis, and 3–6% for an annual basis during occupied hours.

3.4.3. ISO 7730

Metric

The metrics suggested in ISO 7730 [34] are:

(a) Percentage of occupied hours outside the range (*%OhOR*).
(b) Degree hours (*Dh*).
(c) Weighted *PPD* ($\sum$ PPD).
(d) Average *PPD*.
(e) Total *PPD*.

The difference between weighted *PPD* in ISO 7730 compared with EN 15251 is that in ISO 7730, percentage of discomfort happens when *PMV* is $\geq$+0.5 or $-$0.5, whilst in EN15251, percentage of discomfort counts when *PMV* is bigger > than +0.5 or less than $-$0.5. The average *PPD* metric presents the average accumulation of *PPD* during the occupied hours. The total *PPD* calculates the accumulation of all *PPD*s over the occupied hours.

Index

The indices suggested in ISO 7730 are operative temperature, and *PMV*.

Limitations of ISO 7730 are listed below:

(a) There is not any threshold limit for all the defined metrics (*%OhOR*, *Dh*, weighted *PPD*, average *PPD*, and total *PPD*).
(b) The index operative temperature does not consider air movement and relative humidity.

3.4.4. CIBSE Guide A (2006)

Metric

A single metric proposed in CIBSE Guide A (2006) [31] is percentage of occupied hours outside the range (*%OhOR*), which is defined only for the naturally ventilated buildings. It is calculated based on the summation of occupied hours above the threshold during the total number of occupied hours. The threshold limit is defined for up to 1% during the occupied hours over a year in bedrooms and living rooms in residential buildings.

Index

The index proposed in CIBSE Guide A (2006) is operative temperature.

Limitations of CIBSE Guide A are listed below:

(a) The proposed index is operative temperature, which does not consider two environmental parameters, namely relative humidity and indoor air movement; and two personal parameters, known as metabolic rate and clothing value.
(b) It does not include the intensity of exposure to heat.

3.4.5. CIBSE TM52

CIBSE TM52 [16] defines three criteria for the overheating risk assessment for only free-running buildings. Each dwelling that fails to meet any two of the three criteria is assessed as overheated.

CIBSE TM52 Criteria I

Metric

The metric proposed in CIBSE TM52 criteria I is percentage of exceedance hours (%*Eh*).

The %*Eh* is the percentage of occupied hours, when indoor operative temperature exceeds the threshold limit by 1 degree centigrade or above during the total number of occupied hours. This metric requires the calculation over the warm season (1 May to 30 September in Northern Hemisphere). The maximum limit is allowed for 3% during the occupied hours during the summertime.

CIBSE TM52 Criteria II

Metric

In criteria II, CIBSE TM52 focuses on overheating within one day and introduces daily degree exceedance hours (*DEh*). It quantifies the summation of exceedance hours multiplied to the intensity of heat. The intensity of heat is calculated as delta between the operative temperature and the maximum threshold limit. The maximum acceptable daily *DEh* is 6-degree exceedance hours in any one day to limit the overheating in extreme events such as during heatwaves.

CIBSE TM52 Criteria III

This criterion includes a maximum value in which the indoor operative temperature should not be exceeded by 4 degrees Celsius. This criterion does not consider the summation (intensity) of exposure to heat over the time.

Index

The index proposed in CIBSE TM52 for all three criteria is operative temperature.

Limitations for CIBSE TM 52 are listed as follows:

(a) The metric %*Eh* does not include intensity of exposure to heat.
(b) The index, operative temperature does not include the two other environmental parameters (relative humidity, and indoor air movement), and two personal parameters (clothing value, and metabolic rate).

3.4.6. CIBSE TM59

Metric

The proposed metrics in CIBSE TM59 [32]:

(a) For mechanically ventilated buildings, CIBSE TM59 applies a fixed set point, and forcing that operative temperature should not exceed of 26 °C for more than 3% during the occupied hours over a year in all zones (bedroom, kitchen, and living room).
(b) Naturally ventilated buildings need to pass two criteria as follows:
    (1) All zones including bedrooms, living room, and kitchen need to follow criteria I in CIBSE TM52.
    (2) For the bedrooms, an additional metric named percentage of sleeping hours outside the range (%*ShOR*) is introduced. The %*ShOR* dictates that the operative temperature should not exceed 26 °C for more than 1% over a year during the sleeping hours (10:00 p.m.–7:00 a.m.) in the bedrooms.

Index

The index applied in CIBSE TM59 is operative temperature.

Limitation of CIBSE TM59:

(a) The index proposed in TM59 is operative temperature, which does not consider relative humidity, indoor air speed, metabolic rate, and clothing insulation for the comfort calculation.
(b) The intensity of exposure to heat is overlooked in the metrics of TM59.
(c) The maximum threshold limitations are designed for the long-term overheating assessment and does not take into account the short-term overheating and extreme heat events, such as heatwaves.

3.4.7. CIBSE Guide A (2015)

Metric

The metrics applied in CIBSE Guide A (2015):

(a) For the buildings with mixed mode ventilation (combination of mechanical and natural ventilation), CIBSE Guide A (2015) suggests percentage of occupied hours outside the range (*%OhOR*) metric. *%OhOR* shows the percentage of occupied hours above the comfort zone. The calculated *%OhOR* should not exceed 3% during the occupied hours over a year.
(b) For naturally ventilated buildings CIBSE Guide A (2015), follows the three criteria proposed in CIBSE TM52.

Index

The indices used in CIBSE Guide A are operative temperature, and *PMV*.
Limitations for CIBSE Guide A (2015) are listed below:

(a) The metric *%OhOR* does not consider the intensity of exposure to heat.
(b) The indices applied are operative temperature and *PMV*, which do not include two environmental parameters (relative humidity and air speed) in the calculation of comfort.

3.4.8. Passive House

Metric

The metric introduced in Passive House [35] for passively cooled buildings (no active cooling such as air conditioning or HVAC systems) is the percentage of hours outside the range (*%hOR*) for occupied and unoccupied hours. It refers to the percentage of exceedance hours above 25 °C operative temperature. It requires that all zones not exceed 10% over a year.

Index

Index applied in Passive House standard is operative temperature.
Limitations of Passive House criteria:

(a) The metric *%hOR* does not include the intensity of overheating.
(b) The metric does not include the occupied hours.
(c) The index applied is operative temperature, which does not count two environmental parameters of relative humidity and indoor air movement and two personal parameters of metabolic rate and clothing value in the comfort estimation.
(d) It assesses the long-term risk of overheating over a year. Therefore, the short-term overheating assessment is overlooked.

3.4.9. ASHRAE-55 (2017) and ASHRAE-55 (2020)

Metrics

Metrics proposed in ASHRAE 55 (2017) and (2020) are listed below:

(a) Exceedance hours (*Eh*).
(b) Degree Exceedance hours (*DEh*).

Exceedance hours refer to accumulation of number of occupied hours when *PMV* or operative temperature falls beyond the thermal comfort zone threshold. The comfort zone could follow either static comfort zone boundaries or adaptive comfort zone boundaries.

The degree exceedance hours are calculated based on the number of exceedance occupied hours multiplied to the intensity of discomfort. For the static model, the intensity of discomfort is calculated as the accumulation of *PMV*s above +0.5 and less than −0.5. For the adaptive model, degree of discomfort is calculated as the summation of operative temperatures above the upper limit of the comfort zone (for the overheating) and the summation of operative temperatures below the lower limit of comfort zone for the over cooling.

Index

The indices proposed in both ASRAE 55 (2017) and ASHRAE 55 (2020) are operative temperature and *PMV*.

Limitations of ASHRAE 55 (2027) and (2020):

(a) The exceedance hours (*Eh*) metric does not calculate intensity of exposure to heat.
(b) The maximum acceptable limit is not identified.

Table 1 summarises the metrics, indices, threshold limitations, building type, and the cooling operation systems in all the described regulatory documents described above.

Table 1. Analytical and critical comparison of international standards. Avg = average, *%OhOR* = percentage of occupied hours outside the range; T_{oi} = indoor operative temperature; *%hOR* = percentage of hours outside the range; *%ShOR* = percentage of sleeping hours outside the range; *Dh* = degree hours; *Eh* = exceedance hours; NV = natural ventilation; and mechanical ventilation = MV.

Standard	Metrics	Index	Threshold Limit	Thermal Comfort Model	Building Type	Building Operation System
EN 15251[19]	*%OhOR*; *Dh*; weighted *PPD*	*PMV*; T_{oi}	Acceptable deviation of 3–5% of *%OhOR* for daily, weekly, monthly, and yearly periods during the occupied hours.	Static and adaptive model	All building types, with occupants' availability to adaptation response.	MV and NV
EN 16798 [33]	*%OhOR*; *Dh*; weighted *PPD*	*PMV*; T_{oi}	Acceptable deviation of 20–50% of *%OhOR* (weekly); 12–25% (monthly); 3–6% (annually) deviation during occupied hours.	Static and adaptive model	All with occupant availability to adaptation response.	MV and NV
ISO 7730 [34]	*%OhOR*; *Dh*; weighted *PPD*; average *PPD*; total *PPD*	*PMV*	Not specified	Static and adaptive model	All with occupant availability to adaptation response.	MV and NV
ASHRAE 55 [29,30]	Exceedance hours (*Eh*); degree exceedance hours (*DEh*)	*PMV*; T_{oi}	Not specified	Static and adaptive model	All with occupant availability to adaptation response.	MV and NV

Table 1. *Cont.*

Standard	Metrics	Index	Threshold Limit	Thermal Comfort Model	Building Type	Building Operation System
CIBSE Guide A (2006) [31]	*%OhOR*	T_{oi}	Max 1% deviation of *%OhOR* during occupied hours over a year in bedrooms and living room.	NA	Residential buildings	NV
CIBSE TM52 [18]	*%Eh* (Criteria I); *DEh* (Criteria II); T_{oi} (Criteria III).	T_{oi}	Criterion I: max 3% deviation of *%Eh* during occupied hours during a year. Criterion II: max 6 degree exceedance hours during a day. Criterion III: max 4 °C operative temperature right now right here.	NA	Residential buildings	NV
CIBCE Guide A (2015) [31]	*%OhOR,*	*PMV*, T_{oi}	Max 3% on *%OhOR* during the occupied hours for MV buildings; for NV buildings same as CIBSE TM52.	NA	Residential buildings	MV and NV
CIBCE TM59 [32]	*%Eh* *%ShOR*	T_{oi}	For MV buildings: Max T_{oi} of 26 °C for max 3% during the occupied hours over a year in all zones. For NV buildings: (a) living, kitchen, and bedroom follow CIBSE TM 52; and (b) bedrooms max Toi of 26 °C for max 1% of total annual hours during the sleeping hours.	NA	Residential buildings	MV and NV
Passive house [35]	*%hOR*	T_{oi}	Max 10% of *%hOR* when T_{oi} is above 25 °C for all living areas.	NA	All	Passive cooling

3.5. Analytical and Critical Comparison of National Building Codes

The European Commission established a framework including the Energy Performance of Building Directive (EPBD) in 2012, aiming to reduce CO_2 emission and improve energy performance of the buildings by 2050. The member states of the European Commission have to implement the general regulations made by EPBD. This manuscript analysed and compared the thermal comfort and overheating assessment criteria embedded in the national building codes. For each available regulatory document, we extracted the employed standard, index, threshold limit, building type, and the building operation system (see Table 2). Only the available English version of national codes is included in this study.

Table 2. Analytical and critical comparison national building codes based on the Energy Performance of Building Directive (EPBD). NV = natural ventilation; Resi = residential building; T_{ai} = indoor air temperature; MRT_i = indoor mean radiant temperature; NS = not specified; NA = not applicable; IT = internal index (UK); IO = overheating index (Belgium); S_{oh} = solar transmittance index (Germany); and S_{zul} = sum of solar input parameters given in DIN 4108 (Germany).

National Building Code	Standard	Index	Threshold Limit	Thermal Comfort Parameters Included	Occupant's Adaptive Response	Building Type	Building Operation System
Belgium (Brussels)	Passive House	T_{oi}	Max 25 °C by 5% during the occupied hours over a year.	T_{ai}; MRT_i	No	Resi	NV
Belgium (Wallonia and Flanders)	ISO 13790	IO	1000 $Dh < I_{oh} <$ 6500 Dh	No	No	Resi	NV
Germany	DIN 4108	T_{oi}	Compliance through passing one of the criteria. Criteria I: $S_{oh} \leq S_{zul}$ (sum of solar input parameters given in DIN 4108). Criterion II: threshold temperature depends on climatic region (25 °C, 26 °C, and 27 °C), $Dh <$ 1200 Dh during a year.	T_{ai}; MRT_i	N0	Resi	Passive cooling and active cooling
France	EN 15251	T_{oi}	$Dh <$ 1250	T_{ai}; MRT_i	Yes	Resi	NV
UK	SAP	IT	NA	T_{ai}	No	Resi	NV
Netherlands	NS	T_{OJULY}; GTO; *PMV*	Criteria I: max limit value of 1 for T_{OJULY}. Criteria II: max threshold value of 450 for GTO in living area (when *PMV* exceed +0.5).	T_{ai}; MRT_i	No	Resi	NV

3.5.1. France

France adopted the EN 15251 standard and implements the degree hours (*Dh*) metric and indoor operative temperature index to assess discomfort in new residential buildings during the summer months [12]. The threshold for the comfort zone is derived from the upper limit of adaptive comfort model used in EN 15251. The maximum acceptable exceedance operative temperature is 1250 *Dh* for the summertime. It could be translated to a 25-day period of 28 °C indoor operative temperature during the nighttime and 30 °C during the daytime. The weather file for the assessment should be similar to the weather data in 2003, when a heatwave occurred.

Critical review of the criteria:

(a) The criteria are appliable in all climate zones across the country (France has different climate zones, i.e., Oceanic and Mediterranean).
(b) The criteria are for the long-term assessment (during the summertime), therefore, extreme events such as heatwave days could be neglected.
(c) All hours including occupied, and non-occupied hours are included.

(d) The index applied is operative temperature, which excludes four other comfort parameters (i.e., relative humidity, indoor air movement, clothing insulation, and metabolic rate).

3.5.2. Belgium

Belgium is a country with three regions, namely Brussels, Flanders, and Wallonia. Each region implies its own regulation to meet the EPBD requirements. Brussels implements the thermal comfort criteria in residential buildings based on the Passive House standard. The operative temperature during the occupied hours should not exceed 25 °C for maximum of 5% during a year [15].

In Flanders and Wallonia, an overheating index (I_{Oh}) is introduced which is derived from ISO 13790 and is calculated based on the heat balance equation taking into account the heat gain and heat loss of the building in degree hours (*Dh*). The acceptable range of I_{oh} is between 1000 *Dh* and 6500 *Dh*.

Critical review of these criteria includes:

(a) The indices Toi and I_{oh} do not include two environmental parameters (relative humidity, and indoor air speed) nor two personal parameters (clothing value and metabolic rate).
(b) It uses same threshold limitation for short-term and long-term exposure to heat stress.
(c) It counts all the occupied and non-occupied hours.
(d) It does not consider the adaptation responses from occupants.
(e) The intensity of exposure to heat is not counted.

3.5.3. Germany

In Germany, the overheating assessment method includes two criteria, and each building needs to meet at least one criterion to pass the assessment.

Criterion I

This criterion is based on solar heat gains in the most critical room. A solar transmittance index is introduced (S_{trans}) which is the function of window type, window size, room size, and shading systems.

Criterion II

Criterion II proposes the degree hours (*Dh*) metric. The maximum acceptable *Dh* is 1200 *Dh* over a year in the most critical room. The threshold operative temperature depends on the climate zone ranging between 25 and 27 °C.

Critical reviews of these criteria are listed below:

(a) Criterion I is based on solar heat gain, therefore, the heat gain from occupants and devices, i.e., cooking heat, are not included.
(b) Criterion I does not include the intensity of heat exposure.
(c) In criterion I, the environmental and personal parameters of thermal comfort calculation are not included.
(d) In criterion I, overheating assessment is time independent.
(e) Both criteria do not consider the occupants' adaptation responses.
(f) Both criteria are not normalised to occupied hours.

3.5.4. Netherland

In Netherland there are two criteria [12]:

Criteria I, introduce a dimensionless index only for the month of July, named overheating index July (I_{OJuly}). The index is calculated based on the number of external façades of the building and is derived from the average cooling requirement of the entire building. The maximum limit value for I_{OJuly} is 1.

Criterion II is based on weighted *PPD* when *PMV* exceed +0.5. The index is named GTO and has a maximum limitation of 450.

These criteria could be criticised as:

(a) Both criteria are only applied in July. Therefore, other extreme events during other summer months are neglected.

Criteria I is based on heat balance equation and does not include thermal comfort parameters.

Criteria I is for all hours, including occupied and unoccupied hours.

Occupants' adaptation responses are not considered in both criteria.

3.5.5. United Kingdom

In the United Kingdom, the overheating evaluation is based on the standard assessment procedure (SAP) [21]. SAP introduces an index named internal threshold (IT) for overheating assessment during summer months (June, July, and August). IT is applied to estimate the likelihood of high indoor heat. In the calculation of IT, external air temperature, heat gains, heat losses, and building thermal mass are considered.

These criteria could be criticised as:

(a) The criteria are specified for summer months only, therefore, other extreme events outside the summertime could be excluded.
(b) The overheating assessment is for the long-term assessment (monthly), whilst the short-term events would be neglected.
(c) The index is based on air temperature therefore other parameters impacting the thermal comfort are not included (relative humidity, air velocity, personal metabolic rate, and clothing value).
(d) It calculates all hours including occupied hours and non-occupied hours.

Table 2 summarises the overheating assessment criteria in five national building codes.

The analysis of international standards reveals that most of the standards, the employed indices focusing primarily on the air temperature and operative temperature, neglect crucial environmental parameters (relative humidity and indoor air velocity) and personal factors (clothing value and metabolic rate) that significantly influence occupants' thermal sensation [50]. Furthermore, the regulatory documents often overlook occupants' adaptation responses. In naturally ventilated buildings, where occupants are more exposed to outdoor air temperature, the consideration of adaptive thermal comfort becomes pertinent [49]. The thermal comfort threshold and overheating criteria in the regulatory documents play a significant role in the energy consumption or savings in the buildings.

Following reviewing the international and national regulatory documents, we suggest following measures for the improvement or development of regulatory overheating criteria.

(a) Introduce a comprehensive metric: Propose a metric which measures both intensity and duration of exposure to heat stress, such as the degree hours metric, examined by Sadeghi et al. (2021) [8]. During extreme events such as heat waves, the human body's thermal regulation responds not only to the intensity of heat, but also to the cumulative duration of exposure. Prolonged exposure to elevated temperatures can overwhelm the body's ability to dissipate heat, increasing the risk of heat-related illnesses. By integrating both factors, this metric provides a more holistic understanding of heat stress, enabling better assessment of overheating risks in residential buildings.
(b) Thermal comfort index: apply an index that quantifies thermal sensation by encompassing four environmental parameters (indoor air temperature, mean radiant temperature, relative humidity, and indoor air velocity), along with two personal parameters (metabolic rate and clothing insulation).
(c) Incorporate short-term and long-term assessment: implement a nuanced approach to consider both short-term evaluations (i.e., extreme heat events such as heatwaves), and long-term overheating assessment (i.e., over a year or over summertime).

(d) Account for occupant adaptation: Include considerations for occupant adaptation responses, acknowledging that occupants may modify their behaviour based on the thermal conditions. For example, with regards to occupants' clothing value or metabolic rate, it could consider typical summer clothing value for calculations over summertime; or assume a metabolic rate of a standing relaxing person for the kitchen and sitting rooms predictions.

(e) Climate zone differentiation: recognise the diverse climates across the country addressing the unique thermal challenges of each climate zone as classified by the national climate classification, i.e., National Construction Code (NCC) in Australia [23].

4. Conclusions

In this study, we explored and critically compared international standards and national building codes related to the assessment of thermal comfort and overheating risk in residential buildings. The findings reveal that most of the standards employ indices, particularly focusing on air and operative temperatures, yet often overlooking crucial environmental parameters and personal factors that significantly impact occupants' thermal experiences.

International standards such as ASHRAE-55, EN 15251, and ISO 7730 predominantly utilise metrics such as percentage of occupied hours outside the range (*%OhOR*), and weighted predicted percentage dissatisfied (*PPD*). While these metrics offer insights, they exhibit limitations, including a lack of consideration for the intensity of exposure to heat, uniform threshold limitations for short and long-term assessments, and neglect of adaptation responses from occupants.

The static thermal comfort models, exemplified by the predicted mean vote (*PMV*), offer a traditional yet limited approach by assuming uniform preferences among occupants. Conversely, the adaptive thermal comfort model recognises the dynamic nature of indoor environments, individual preferences, and adaptation responses, providing a more nuanced and energy-efficient framework especially suitable for naturally ventilated buildings [49].

Analysing national building codes, we explored criteria from France, Belgium, Germany, Netherlands, and the United Kingdom. Notably, these criteria vary in their approaches, introducing indices such as internal threshold (IT), overheating index (IO), and overheating metrics for specific months. However, common limitations persist, such as overlooking short-term events, neglecting occupants' adaptation responses, and focusing predominantly on the air temperature.

Considering the absence of mandatory regulations for overheating criteria in residential buildings in some countries across the world, i.e., Australia, we propose a comprehensive approach for developing new overheating regulatory documents specified to the local climate. Suggestions include introducing a metric to measure both intensity and duration of heat stress, employing a thermal comfort index encompassing environmental and personal parameters, differentiating between short-term and long-term assessments, accounting for occupant adaptation, and recognising the diverse climate zones across the country.

This study faced certain limitations that should be addressed in future research. Some national regulatory documents, such as those from Italy and Greece, were only available in local languages, and their English versions were inaccessible. As a result, the analysis relied on European legislation and documentation, limiting the inclusion of region-specific details from these countries. Future studies may look at possibilities for developing climate-adaptive regulations tailored to the needs of individual countries, which will be crucial to addressing the unique challenges posed by varying geographic and climatic conditions.

Addressing overheating risk in buildings requires a holistic and adaptive approach that considers the diverse factors influencing thermal comfort. Striking knowledge of human thermal physiology, human adaptation responses, building physics, and environmental climate especially in geographically varied countries such as Australia, is crucial for creating effective and sustainable regulations to ensure occupants' well-being and energy-efficient building practices. Our findings have implications for policymakers, building energy assessors, building designers, and public health professionals.

Author Contributions: Conceptualization, M.S., D.C. and A.W.; methodology, M.S. and D.C.; software, M.S.; validation, M.S. and D.C.; formal analysis, M.S.; investigation, M.S.; resources, D.C. and. A.W.; data curation, M.S.; writing—original draft preparation, M.S.; writing—review and editing, M.S., D.C. and A.W.; visualization, M.S.; supervision, D.C. and A.W.; project administration, D.C. and A.W.; funding acquisition, D.C. and A.W. All authors have read and agreed to the published version of the manuscript.

Funding: This research was funded by Commonwealth Scientific Industrial Research Organisation (CSIRO). Information regarding the funder and the funding number should be provided. Please check the accuracy of funding data and any other information carefully.

Acknowledgments: This study has been funded by Commonwealth Scientific Industrial Research Organisation (CSIRO). The project has been led and conducted by CSIRO's Building Energy Efficiency team within Energy Research Unit.

Conflicts of Interest: The authors declare no conflict of interest.

Nomenclature

T_{oi}	Operative temperature—indoor (°C)
MRT	Mean radiant temperature—indoor (°C)
T_{ai}	Air temperature—indoor (°C)
T_{aout}	Air temperature—outdoor (°C)
PMV	Predicted mean vote (dimensionless)
PPD	Predicted percentage dissatisfied (%)
h	Hour (h)
Oh	Occupied hours (h)
Dh	Degree hours (°Ch)
Eh	Exceedance hours (h)
Met	Metabolic rate (met)
Clo	Clothing value (clo)
V_i	Air speed—indoor (m/s)
W	Wind speed—outdoor (m/s)
$Tarm_{out}$	Running mean air temperature—outdoor (°C)
T_n	Neutral air temperature (°C)
T_{accept}	Acceptable air temperature (°C)
T_{com}	Comfort temperature (°C)
%OhOR	Percentage of occupied hours outside the range (%)
%hOR	Percentage of hours outside the range (%)
%ShOR	Percentage of sleeping hours outside the range (%)

References

1. Hamdy, M.; Carlucci, S.; Hoes, P.-J.; Hensen, J.L.M. The impact of climate change on the overheating risk in dwellings—A Dutch case study. *Build. Environ.* **2017**, *122*, 307–323. [CrossRef]
2. Jay, O.; Capon, A.; Berry, P.; Broderick, C.; de Dear, R.; Havenith, G.; Honda, Y.; Kovats, R.S.; Ma, W.; Malik, A.; et al. Reducing the health effects of hot weather and heat extremes: From personal cooling strategies to green cities. *Lancet* **2021**, *398*, 709–724. [CrossRef] [PubMed]
3. Ebi, K.L.; Capon, A.; Berry, P.; Broderick, C.; de Dear, R.; Havenith, G.; Honda, Y.; Kovats, R.S.; Ma, W.; Malik, A. Hot weather and heat extremes: Health risks. *Lancet* **2021**, *398*, 698–708. [CrossRef] [PubMed]
4. Santamouris, M. Recent progress on urban overheating and heat island research. Integrated assessment of the energy, environmental, vulnerability and health impact. Synergies with the global climate change. *Energy Build.* **2020**, *207*, 109482. [CrossRef]
5. Gasparrini, A.; Guo, Y.; Sera, F.; Vicedo-Cabrera, A.M.; Huber, V.; Tong, S.; de Sousa Zanotti Stagliorio Coelho, M.; Nascimento Saldiva, P.H.; Lavigne, E.; Correa, P.M.; et al. Projections of temperature-related excess mortality under climate change scenarios. *Lancet Planet. Health* **2017**, *1*, e360–e367. [CrossRef]
6. Sadeghi, M.; Chaston, T.; Hanigan, I.; de Dear, R.; Santamouris, M.; Jalaludin, B.; Morgan, G.G. The health benefits of greening strategies to cool urban environments—A heat health impact method. *Build. Environ.* **2022**, *207*, 108546. [CrossRef]

7. Ide, T. Climate change and Australia's national security. *Aust. J. Int. Aff.* **2023**, *77*, 26–44. [CrossRef]
8. Sadeghi, M.; de Dear, R.; Morgan, G.; Santamouris, M.; Jalaludin, B. Development of a heat stress exposure metric—Impact of intensity and duration of exposure to heat on physiological thermal regulation. *Build. Environ.* **2021**, *200*, 107947. [CrossRef]
9. Nematchoua, M.K.; Sadeghi, M.; Reiter, S.; Attia, S. Roadmap to reduce the direct effects of climate change on building performance; A case study applied to the top 8 deserts and top 8 coldest regions in the world. *Next Sustain.* **2023**, *1*, 100007. [CrossRef]
10. Livada, I.; Pyrgou, A.; Haddad, S.; Sadeghi, M.; Santamouris, M. Recent Climatic Trends and Analysis of Monthly Heating and Cooling Degree Hours in Sydney. *Climate* **2021**, *9*, 114. [CrossRef]
11. Nematchoua, M.K.; Sadeghi, M.; Reiter, S. Strategies and scenarios to reduce energy consumption and CO_2 emission in the urban, rural and sustainable neighbourhoods. *Sustain. Cities Soc.* **2021**, *72*, 103053. [CrossRef]
12. Beggs, P.J.; Zhang, Y.; McGushin, A.; Trueck, S.; Linnenluecke, M.K.; Bambrick, H.; Capon, A.G.; Vardoulakis, S.; Green, D.; Malik, A. The 2022 report of the MJA–Lancet Countdown on health and climate change: Australia unprepared and paying the price. *Med. J. Aust.* **2022**, *217*, 439–458. [CrossRef] [PubMed]
13. Di Napoli, C.; Pappenberger, F.; Cloke, H.L. Assessing heat-related health risk in Europe via the Universal Thermal Climate Index (UTCI). *Int. J. Biometeorol.* **2018**, *62*, 1155–1165. [CrossRef]
14. Rahif, R.; Amaripadath, D.; Attia, S. Review on Time-Integrated Overheating Evaluation Methods for Residential Buildings in Temperate Climates of Europe. *Energy Build.* **2021**, *252*, 111463. [CrossRef]
15. *BS EN 15251*; Indoor Environmental Input Parameters for Design and Assessment of Energy Performance of Buildings Addressing Indoor Air Quality, Thermal Environment, Lighting and Acoustics. British Standards Institution: London, UK, 2007.
16. NHBC. *Overheating in New Homes—A Review of the Evidence*; National House Building Council Foundation: Bucks, UK, 2012.
17. United Nations. *United Nations Economic Commission for Europe, Best Policy Practice for Promoting Energy Efficiency*; United Nations: New York, NY, USA; Geneva, Switzerland, 2015. Available online: https://unece.org/DAM/energy/se/pdfs/geee/pub/ECE_Best_Practices_in_EE_publication.pdf (accessed on 10 June 2024).
18. The Chartered Institution of Building Services Engineers (CIBSE) TM52. *The Limits of Thermal Comfort: Avoiding Overheating in European Buildings*; The Chartered Institution of Building Services Engineers: London, UK, 2013.
19. *EN 15251:2007*; Indoor Environmental Input Parameters for Design and Assessment of Energy Performance of Buildings Addressing Indoor Air Quality, Thermal Environment, Lighting and Acoustics. European Committee for Standardization: Brussels, Belgium, 2007.
20. Attia, S.; Benzidane, C.; Rahif, R.; Amaripadath, D.; Hamdy, M.; Holzer, P.; Koch, A.; Maas, A.; Moosberger, S.; Petersen, S.; et al. Overheating calculation methods, criteria, and indicators in European regulation for residential buildings. *Energy Build.* **2023**, *292*, 113170. [CrossRef]
21. Nicol, F. *TM52: The Limits of Thermal Comfort: Avoiding Overheating in European Buildings*; CIBSE: London, UK, 2013; Volume 1.
22. Kim, J.; Xiong, J.; de Dear, R.; Parkinson, T.; Jeong, B.; Wu, Z.; Sadeghi, M.; Chen, D. Testing the applicability of CIBSE overheating criteria to Australian subtropical residential contexts. *Build. Environ.* **2023**, *246*, 110987. [CrossRef]
23. Carlucci, S.; Pagliano, L. A review of indices for the long-term evaluation of the general thermal comfort conditions in buildings. *Energy Build.* **2012**, *53*, 194–205. [CrossRef]
24. Salem, R.; Bahadori-Jahromi, A.; Mylona, A. Investigating the impacts of a changing climate on the risk of overheating and energy performance for a UK retirement village adapted to the nZEB standards. *Build. Serv. Eng. Res. Technol.* **2019**, *40*, 470–491. [CrossRef]
25. *ASHRAE Standard 55-2023*; Thermal Environmental Conditions for Human Occupancy. ASHRAE: Atlanta, GA, USA, 2023.
26. Guo, H.; Ferrara, M.; Coleman, J.; Loyola, M.; Meggers, F. Simulation and measurement of air temperatures and mean radiant temperatures in a radiantly heated indoor space. *Energy* **2020**, *193*, 116369. [CrossRef]
27. Australia Building Codes Board (ABCB). *National Construction Code (NCC) 2019 Volume One*; ABCB: Canberra, Australia, 2019.
28. Esri 10.7.1. "Topographic" [Basemap], World Topographic Map, 2019. Available online: https://www.arcgis.com/home/item.html?id=30e5fe3149c34df1ba922e6f5bbf808f (accessed on 22 September 2022).
29. *ANSI/ASHRAE Standard 55-2017*; Thermal Environmental Conditions for Human Occupancy. American Society of Heating, Refrigerating and Air Conditioning Engineers: Atlanta, GA, USA, 2017.
30. Vaughn, M. ASHRAE Research Report: 2020–2021. *ASHRAE J.* **2021**, *63*, 73–87.
31. The Chartered Institute of Building Services Engineers (CIBSE). *Guide, A. Environmental Design*; The Chartered Institute of Building Services Engineers (CIBSE): London, UK, 2006.
32. The Chartered Institution of Building Services Engineers (CIBSE) TM59. *Design Methodology for the Assessment of Overheating Risk in Homes*; The Chartered Institution of Building Services Engineers: London, UK, 2017.
33. Khovalyg, D.; Kazanci, O.B.; Halvorsen, H.; Gundlach, I.; Bahnfleth, W.P.; Toftum, J.; Olesen, B.W. Critical review of standards for indoor thermal environment and air quality. *Energy Build.* **2020**, *213*, 109819. [CrossRef]
34. *ISO 7730*; Ergonomics of the Thermal Environment—Analytical Determination and Interpretation of Thermal Comfort Using Calculation of the *PMV* and *PPD* Indices and Local Thermal Comfort Criteria. International Standards Organization: Geneva, Switzerland, 2005; Volume 52.
35. Fletcher, M.J.; Johnston, D.K.; Glew, D.W.; Parker, J.M. An empirical evaluation of temporal overheating in an assisted living Passivhaus dwelling in the UK. *Build. Environ.* **2017**, *121*, 106–118. [CrossRef]

36. Matzarakis, A.; Rutz, F.; Mayer, H. Modelling radiation fluxes in simple and complex environments: Basics of the RayMan model. *Int. J. Biometeorol.* **2010**, *54*, 131–139. [CrossRef] [PubMed]
37. Sadeghi, M.; de Dear, R.; Wood, G.; Samali, B. Development of a bioclimatic wind rose tool for assessment of comfort wind resources in Sydney, Australia for 2013 and 2030. *Int. J. Biometeorol.* **2018**, *62*, 1963–1972. [CrossRef] [PubMed]
38. Sadeghi, M. Thermal Comfort Performance of Wind Towers in the Australian Residential Context. The University of Sydney. 2017. Available online: http://refhub.elsevier.com/S0360-1323(21)00939-2/sref33 (accessed on 20 July 2024).
39. Sadeghi, M. The Impact of Greenery on Heat-Related Mortality in Sydney, Australia. In *Urban Overheating: Heat Mitigation and the Impact on Health*; Aghamohammadi, N., Santamouris, M., Eds.; Springer Nature: Singapore, 2022; pp. 181–195. [CrossRef]
40. Gagge, A.P.; Stolwijk, J.A.J.; Hardy, J.D. Comfort and thermal sensations and associated physiological responses at various ambient temperatures. *Environ. Res.* **1967**, *1*, 1–20. [CrossRef]
41. Sadeghi, M.; de Dear, R.; Samali, B.; Wood, G. Application of wind towers in the Australian residential context–a wind tunnel assessment of thermal comfort performance. In Proceedings of the 9th Windsor Conference, Windsor, UK, 7–10 April 2016.
42. Sadeghi, M.; Wood, G.; Samali, B.; de Dear, R. Effects of urban context on the indoor thermal comfort performance of windcatchers in a residential setting. *Energy Build.* **2020**, *219*, 110010. [CrossRef]
43. Van Hoof, J. Forty years of Fanger's model of thermal comfort: Comfort for all? *Indoor Air* **2008**, *18*, 182–201. [CrossRef] [PubMed]
44. Fanger, P. Thermal comfort: Analysis and applications in environmental engineering. *Perspect Public Health* **1970**, *92*, 164.
45. Cheung, T.; Schiavon, S.; Parkinson, T.; Li, P.; Brager, G. Analysis of the accuracy on *PMV–PPD* model using the ASHRAE Global Thermal Comfort Database II. *Build. Environ.* **2019**, *153*, 205–217. [CrossRef]
46. Griffiths, I. *Thermal Comfort Studies in Buildings with Passive Solar Features, Field Studies;* Report to the Commission of the European Community; Commission of the European Communities: Brussels, Belgium, 1990; Volume 35.
47. Nicol, J.F.; Hacker, J.; Spires, B.; Davies, H. Suggestion for new approach to overheating diagnostics. *Build. Res. Inf.* **2009**, *37*, 348–357. [CrossRef]
48. Nicol, F. Adaptive thermal comfort standards in the hot–humid tropics. *Energy Build.* **2004**, *36*, 628–637. [CrossRef]
49. De Dear, R.; Brager, G.S. Developing an adaptive model of thermal comfort and preference. *ASHRAE Trans.* **1998**, *104*, 145–167.
50. Sadeghi, M.; Williamson, T.; Miller, W.; Kieboom, J.; Chen, D. Development of a Comfort Rating Method for Australia's Nationwide House Energy Rating Scheme (NatHERS)–Darwin Houses Case Study. In Proceedings of the CATE 2023, Ahmedabad, India, 13–15 December 2023.

energies

MDPI

Article

Concentration of CO_2 in the Local Working Area during the Joint Operation of a Gas Infrared Heater and an Air-Exchange System

Boris Vladimirovich Borisov, Geniy Vladimirovich Kuznetsov, Vyacheslav Ivanovich Maksimov *, Tatiana Aleksandrovna Nagornova and Felix Yurievich Salikhov

School of Energy and Power Engineering, National Research Tomsk Polytechnic University, 634050 Tomsk, Russia
* Correspondence: elf@tpu.ru

Abstract: The formation of local areas in large buildings with regulated thermal conditions is an urgent task. The use of gas infrared heaters for this purpose raises questions on the utility of an air-exchange system and the monitoring of the combustion product concentration. In this study, the modeling of heat transfer processes on premises with a gas infrared heater and an air-exchange system was conducted. The carbon dioxide concentration in the local working area when using a light-type gas infrared heater was determined. The regularities of current formation for circulating air and combustion products on the premises at various air-exchange rates were analyzed. The profiles of CO_2 temperatures and concentrations in the local working areas on the left and right of the equipment model are shown. The article makes a conclusion about the influence of air velocity from the air-exchange system based on average values of carbon dioxide concentration on the premises and in the local working area. The possibility of increasing the temperature in the local working area without exceeding the permissible CO_2 concentrations (less than 1000 ppm) has been identified. The formulated approach allows us to predict the available modes of the air-exchange system to create the highest possible comfort heating parameters while maintaining an acceptable degree of air pollution from combustion products.

Keywords: gas infrared heater; air-exchange system; local working areas; fields of temperatures; field CO_2 concentrations

1. Introduction

The use of gas infrared heaters (GIHs) has a number of advantages over traditional heating systems [1,2]. One of the main benefits lies in the possibility of creating local thermal areas in large industrial premises [3–5]. The creation of such areas is most expedient with the help of light-type GIHs [4,6]. They have a higher power (compared to dark-type ones [4,6]), which is reached through open gas combustion. However, such combustion leads to combustion product emissions (mainly CO_2) into the environment. For this reason, the air-exchange system's operation is required during light-type GIH use [7–9]. When creating thermal areas for a worker, it is advisable to maintain regulated thermal conditions (microclimate) [10–13]. Therefore, parameters such as temperature, air velocity movement, and air composition should be varied within the specified limits. According to standards, the time-weighted average CO_2 in the air composition for premises where people are located should not exceed 1000 ppm (Table 1) [10–13].

Modern methods for calculating gas concentrations in premises and maintaining specified air parameters are mostly based on balance models [10,14,15]. Here, the gas is considered to be evenly distributed on the premises. The CO_2 concentration is measured by sensors installed in arbitrarily selected zones. Based on the data obtained, the mode of air-exchange system operation is controlled, and the regulated air composition is maintained.

Table 1. Indoor air quality classification (EN 13779:2007, GOST 30494-2011) [10–13].

Class	Indoor Air Quality		Acceptable CO_2 Level *, ppm
	Optimal	Permissible	
1	High		400 and less
2	Average		400–600
3		Permissible	600–1000
4		Low	1000 and more

* Allowable CO_2 content in the room is taken over the CO_2 content in the outdoor air outside the room, ppm.

However, recent studies have shown that CO_2 concentrations in the local working area on premises with an operating air-exchange system may not correspond to the average concentrations of pollutants [14,15]. The spatial distribution of carbon dioxide on industrial premises is influenced by many factors: the heat load distribution, the number and location of air inlet and outlet ducts of air-exchange systems, the flow rate and temperature of the supply air, and the location of the CO_2 emission source [16,17]. For example, it was experimentally established that the heat load distribution has a significant effect on the CO_2 distribution in the room [17]. The heated surface intensifies the air movement around it. As a result, the carbon dioxide concentration in this local area decreases. This effect has also been confirmed in numerical simulations [18]. It was found that the heated surface intensified the mixing of air in the local area while reducing the value of the CO_2 concentration by 4.5%. At the same time, the efficiency of air circulation increases by 26.9%.

Heated surfaces in the local working area can significantly affect thermal and concentration fields during the joint operation of open-type gas infrared heaters (CO_2 source) and the air-exchange system [19]. It is also important to mention the established effect of the additional air heating of the local working area [20]. This effect occurs due to the influx of heat released by the GIH case into the air layers around it and their movement into the local working area during air-exchange system operation. It has been experimentally proven that the flow of "fresh" air from the air-exchange system, flowing around the heater case, entrains the gaseous combustion products coming from the GIH burners [20]. As a result, the air-exchange system supplies a mixture of "fresh" air and combustion products to the local working area. This can significantly worsen the air quality in this area. In case of increasing cold airflow from the air-exchange system, the air temperature in the local working area will be significantly reduced. In this regard, it is relevant to analyze the composition of the mixture of gases entering the local working area due to the operation of the gas infrared heater and the air-exchange system.

It Is also reasonable to control t"e airflow in the air-exchange system in order to minimize the gas combustion products influx into the local working area and achieve the maximum possible (within the limits) air temperature in this area. At the same time, the concentration of CO_2 in it will change. However, the analysis and assessment of such changes have not been carried out yet.

The aim of this work was to perform a theoretical analysis of the effect of air-exchange system operation on the thermal conditions and carbon dioxide concentrations in the local working area when using a light-type GIH for the formation of regulated thermal conditions of local working areas.

2. Mathematical Statement of the Problem and Solution Method

Mathematical modeling was carried out within the framework of a two-dimensional approximation for the system shown in Figure 1. A closed rectangular area filled with air was considered, and there was a gas infrared heater (1) and a horizontal panel (simulating equipment) (2) placed in it.

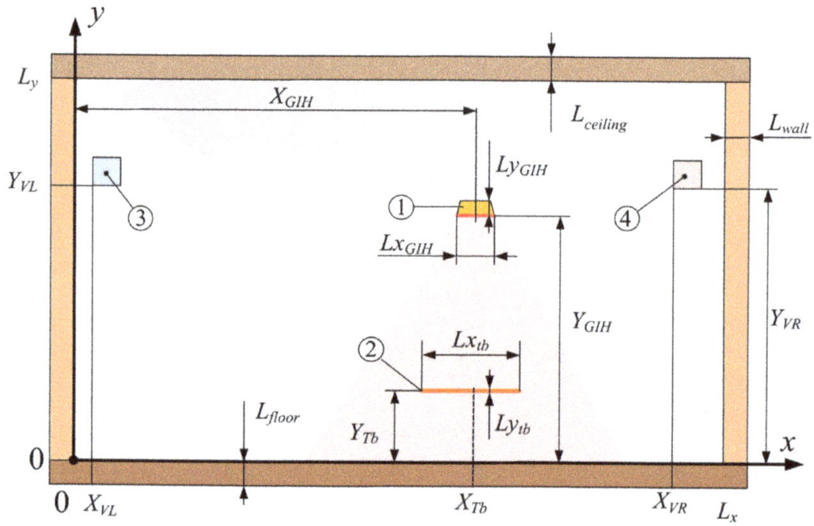

Figure 1. Problem solution area: 1—GIH, 2—panel, 3—air inlet area, 4—air outlet area.

The analysis area also contains the inlet and outlet ducts of the air-exchange system (3 and 4), with the coordinates of the lower left corner $X_{VENT\ left}$, $Y_{VENT\ left}$ and $X_{VENT\ right}$, $Y_{VENT\ right}$. The mathematical model of the considered process involves the creation of the following geometric sets:

1. Total premises volume $A_{\Sigma\ V}$.

$$A_{\Sigma\ V} = \left\{(x,y)\middle| -L_{wall} \le x \le L_x + L_{wall}, -L_{floor} \le y \le L_y + L_{ceiling,}\right\};$$

2. Internal premises volume $A_{\Sigma V\ in}$.

$$A_{\Sigma V\ in} = \{(x,y)|0 \le x \le L_x, 0 \le y \le L_y\};$$

3. Exterior surfaces of the premises $A_{\Sigma F\ out}$.

$$A_{\Sigma F\ out} = \left\{ (x,y) \middle| \begin{array}{c} x = -L_{wall},\ -L_{floor} \le y \le L_y + L_{ceiling} \\ x = L_x + L_{wall},\ -L_{floor} \le y \le L_y + L_{ceiling} \\ -L_{wall} \le x \le L_x + L_{wall},\ y = -L_{floor} \\ -L_{wall} \le x \le L_x + L_{wall},\ y = L_y + L_{ceiling} \end{array} \right\};$$

4. Interior surfaces of the premises $A_{\Sigma F\ in}$.

$$A_{\Sigma F\ in} = \left\{ (x,y) \middle| \begin{array}{l} x = 0,\ \ 0 \le y \le L_y \\ x = L_x,\ \ 0 \le y \le L_y \\ 0 \le x \le L_x,\ \ y = 0 \\ 0 \le x \le L_x,\ \ y = L_y \end{array} \right\};$$

5. Volume of GIH $A_{GIH\ V}$.

$$A_{GIH\ V} = \left\{(x,y)\middle| X_{GIH} - \frac{Lx_{GIH}}{2} \le x \le X_{GIH} + \frac{Lx_{GIH}}{2}, Y_{GIH} \le y \le Y_{GIH} + Ly_{GIH}\right\};$$

6. Total GIH surface $A_{\text{GIH } F}$.

$$A_{\text{GIH } F} = \left\{ (x,y) \middle| \begin{array}{c} x = X_{GIH} - \frac{Lx_{GIH}}{2}, \quad Y_{GIH} \le y \le Y_{GIH} + Ly_{GIH} \\ x = X_{GIH} + \frac{Lx_{GIH}}{2}, \quad Y_{GIH} \le y \le Y_{GIH} + Ly_{GIH} \\ X_{GIH} - \frac{Lx_{GIH}}{2} \le x \le X_{GIH} + \frac{Lx_{GIH}}{2}, \quad y = Y_{GIH} + Ly_{GIH} \\ X_{GIH} - \frac{Lx_{GIH}}{2} \le x \le X_{GIH} + \frac{Lx_{GIH}}{2}, \quad y = Y_{GIH} \end{array} \right\};$$

7. Volume of the horizontal panel (table) $A_{\text{Tb } V}$.

$$A_{\text{Tb } V} = \left\{ (x,y) \middle| X_{Tb} - \frac{Lx_{Tb}}{2} \le x \le X_{Tb} + \frac{Lx_{Tb}}{2}, Y_{Tb} - Ly_{Tb} \le y \le Y_{Tb} \right\};$$

8. Total surface of the horizontal panel (table) $A_{\text{Tb } F}$.

$$A_{\text{Tb } F} = \left\{ (x,y) \middle| \begin{array}{c} x = X_{Tb} - \frac{Lx_{Tb}}{2}, \quad Y_{Tb} - \frac{Ly_{Tb}}{2} \le y \le Y_{Tb} \\ x = X_{Tb} + \frac{Lx_{Tb}}{2}, \quad Y_{Tb} - \frac{Ly_{Tb}}{2} \le y \le Y_{Tb} \\ X_{Tb} - \frac{Lx_{Tb}}{2} \le x \le X_{Tb} + \frac{Lx_{Tb}}{2}, \quad y = Y_{Tb} - Ly_{Tb} \\ X_{Tb} - \frac{Lx_{Tb}}{2} \le x \le X_{Tb} + \frac{Lx_{Tb}}{2}, \quad y = Y_{Tb} \end{array} \right\};$$

9. Volume of the air ducts $A_{\text{VENT } V}$.

$$A_{\text{VENT } V} = \left\{ (x,y) \middle| \begin{array}{c} X_{VENTleft} \le x \le X_{VENTleft} + \frac{Lx_{VENT}}{2}, Y_{VENTleft} \le y \le Y_{VENTleft} + Ly_{VENT} \\ X_{VENTright} \le x \le X_{VENTright} + \frac{Lx_{VENT}}{2}, Y_{VENTright} \le y \le Y_{VENTright} + Ly_{VENT} \end{array} \right\};$$

10. Exterior surface of air ducts $A_{\text{VENT } F}$.

$$A_{\text{VENT } F} = \left\{ (x,y) \middle| \begin{array}{c} X_{VENTleft} \le x \le X_{VENTleft} + Lx_{VENT}, \;\; y = Y_{VENTleft} \\ X_{VENTleft} \le x \le X_{VENTleft} + Lx_{VENT}, \;\; y = Y_{VENTleft} + Ly_{VENT} \\ x = X_{VENTleft}, Y_{VENTleft} \le y \le Y_{VVENTleft} + Ly_{VENT} \\ x = X_{VENTleft} + Lx_{VENT}, Y_{VENTleft} \le y \le Y_{VVENTleft} + Ly_{VENT} \\ X_{VENTright} \le x \le X_{VENTright} + Lx_{VENT}, \;\; y = Y_{VENTright} \\ X_{VENTright} \le x \le X_{VENTright} + Lx_{VENT}, \;\; y = Y_{VENTright} + Ly_{VENT} \\ x = X_{VENTright}, Y_{VENTrigh} \le y \le Y_{VENTrigh} + Ly_{VENT} \\ x = X_{VENTrigh} + Lx_{VENT}, Y_{VENTrigh} \le y \le Y_{VENTrigh} + Ly_{VENT} \end{array} \right\}.$$

Inlet and outlet duct areas $A_{\text{VENT } F\, in}$ and $A_{\text{VENT } F\, out}$.

The convective–conductive heat transfer within the assumed physical model was described by the energy equation [21]:

$$\rho c_P \frac{\partial T}{\partial \tau} + \rho c_P \left(\vec{u} \cdot \nabla \right) T = \nabla \cdot (\lambda \cdot \nabla T), \tag{1}$$

$$(x,y) \in A_{\Sigma\, V} \backslash A_{\text{GIH } V} \backslash A_{\text{VENT } V}$$

where $\tau, \rho, T, c_P, \lambda$—time, density, temperature, specific isobaric heat and thermal conductivity, respectively.

The velocity vector field $\vec{u}$ was determined from the solution of the system of equations of motion and the continuity of an incompressible gas in the Boussinesq approximation [22]:

$$\rho \frac{\partial \vec{u}}{\partial \tau} + \rho \left(\vec{u} \cdot \nabla \right) \vec{u} = \nabla \cdot \left[-p \vec{I} + \vec{K} \right] + (\rho - \rho_0) \vec{g}, \tag{2}$$

$$\rho \frac{\partial \vec{u}}{\partial \tau} + \nabla \cdot \left(\rho \vec{u} \right) = 0, \tag{3}$$

$$(x,y) \in A_{\Sigma\, V\, in} \backslash A_{\text{Tb } V} \backslash A_{\text{GIH } V} \backslash A_{\text{VENT } V}.$$

where $p, \vec{I}$—pressure and unit tensor symbol; $\rho_0, \vec{g}$—initial density and gravitational acceleration;

$\vec{K} = (\mu + \mu_T)\left[\nabla \cdot \vec{u} + \left(\nabla \cdot \vec{u}\right)^T\right] - \frac{2}{3}(\mu + \mu_T)\left(\nabla \cdot \vec{u}\right)\vec{I} - \frac{2}{3}\rho k \vec{I}$—viscous friction stress tensor with allowance for the turbulent component ("T" index), μ—dynamic viscosity coefficient.

The "k-ε" model was used for modeling the turbulent airflow. The turbulence kinetic energy (k) and dissipation rate (ε) are described by the equations [23,24]:

$$\rho\frac{\partial k}{\partial \tau} + \rho\left(\vec{u} \cdot \nabla\right)k = \nabla \cdot \left[\left(\mu + \frac{\mu_T}{\sigma_T}\right)(\nabla \cdot k)\right] + P_k - \rho\varepsilon, \tag{4}$$

$$\rho\frac{\partial \varepsilon}{\partial \tau} + \rho\left(\vec{u} \cdot \nabla\right)\varepsilon = \nabla \cdot \left[\left(\mu + \frac{\mu_T}{\sigma_\varepsilon}\right)(\nabla \cdot k)\varepsilon\right] + C_{\varepsilon 1}\frac{\varepsilon}{k}P_k - C_{\varepsilon 2}\rho\frac{\varepsilon^2}{k}, \tag{5}$$

$$(x, y) \in A_{\Sigma\ V\ in} \backslash A_{\mathrm{Tb}\ V} \backslash A_{\mathrm{GIH}\ V} \backslash A_{\mathrm{VENT}\ V}.$$

Solutions for Equations (4) and (5) were used to calculate $\mu_T = \rho C_\mu \frac{k^2}{\varepsilon}$. In Equations (4) and (5), the operator had the form $P_k = \mu_T\left[\nabla \cdot \vec{u} : \left(\nabla \cdot \vec{u} + \left(\nabla \cdot \vec{u}\right)^T\right) - \frac{2}{3}\left(\nabla \cdot \vec{u}\right)^2\right] - \frac{2}{3}\rho k \nabla \cdot \vec{u}$. The values of the constants were taken according to the general theory [23,24]:

$$C_{\varepsilon 1} = 1.44, C_{\varepsilon 2} = 1.92, C_\mu = 0.09, \sigma_k = 1, \sigma_\varepsilon = 1.3$$

The generation of turbulence kinetic energy in the region under consideration is concentrated in the area of significant temperature and velocity gradients. Such gradients are localized in the area of forced air injection by the air-exchange system and a limited volume above the GIH. This mechanism of turbulence generation is sufficiently correctly described by the chosen model. As further calculations showed, the influence of turbulence on the intensity of mass, momentum, and energy transfer is significant in these same areas. Also, it has a slight effect on the temperature and concentration fields in the rest of the room, even with significant variations in the parameters affecting the turbulent motion intensity.

Radiation fluxes were calculated using the zonal model [25,26], with a direct integration of fluxes between all components ("Surface-to-Surface Radiation") of a closed system of surfaces with angular coefficients determined within this system.

The carbon dioxide (CO_2) supply was assumed to be carried out from the upper boundary of the GIH. There was no pollution generation within the volume of the solution region. Temperature gradients in the region of analysis were relatively small. Thermal diffusion as a process of the second order of significance can be neglected. Fick's Law was used as a diffusion model for the binary "air-carbon dioxide" diffusion process. The main equation for mass conservation in a non-conservative form, and additionally used relations can be written as follows [21,27]:

$$\rho\frac{\partial \omega_i}{\partial \tau} + \rho\left(\vec{u} \cdot \nabla\right)\omega_i = -\nabla \cdot j_i, \tag{6}$$

$$(x, y) \in A_{\Sigma\ V\ in} \backslash A_{\mathrm{Tb}\ V} \backslash A_{\mathrm{GIH}\ V} \backslash A_{\mathrm{VENT}\ V}.$$

$$j_i = -\left(\rho D_i^f \nabla\omega_i + \rho D_i^f \omega_i \frac{\nabla M_n}{M_n} - j_{c,i}\right), \tag{7}$$

$$M_n = \left(\sum_i \frac{\omega_i}{M_i}\right)^{-1}, \ \omega_i \cdot M_n = \chi_i \cdot M_i, \tag{8}$$

$$j_{c,i} = \rho\omega_i \sum_k \frac{M_i}{M_k} D_i^f \nabla\chi_k. \tag{9}$$

The system of Equations (6)–(9) was based on the mass and molar concentrations (ω_i, χ_i), molar masses (M_n) and binary diffusion coefficients (D_i^f). The dependence of the diffusion coefficient D_i^f on temperature was taken into account in the form:

$$D_i^f = D_{i0}^f \left(\frac{T}{T_0} \right)^{1.70} \tag{10}$$

The values of temperatures T_0, zero values of the air velocity components, and the initial value of the mass fraction of CO_2 over the entire area were taken as the initial conditions:

$$T(0,x,y) = T_0, \vec{u}(0,x,y) = 0, \omega_{CO2}(0,x,y) = \omega_{CO2_0},$$
$$(x,y) \in A_{\Sigma\ V} \backslash A_{\mathrm{GIH}\ V} \backslash A_{\mathrm{VENT}\ V}.$$

The temperature of the radiating surface of the GIH was set constant on its lower surface for the entire operation time:

$$T(\tau,x,y) = T_{GIH}, \tau > 0,\ X_{GIH} - \frac{Lx_{GIH}}{2} \le x \le X_{GIH} + \frac{Lx_{GIH}}{2},\ y = Y_{GIH}.$$

During the limited time of GIH operation (work shift), the enclosing structures, as a rule, do not have time to warm up over the entire thickness. Therefore, the conditions of the adiabaticity at the outer boundaries of the solution area were used as boundary conditions for Equation (1):

$$\vec{\nabla T}(\tau,x,y) = 0, \tau > 0,\ (x,y) \in A_{\Sigma\ F\ out}.$$

On the side surfaces of the GIH:

$$\vec{\nabla T}(\tau,x,y) = T_{F\ GIH},\ \tau > 0,$$
$$(x,y) \in \left(x = X_{GIH} - \frac{Lx_{GIH}}{2}, Y_{GIH} \le y \le Y_{GIH} + Ly_{GIH} \right) \cup$$
$$\cup \left(x = X_{GIH} + \frac{Lx_{GIH}}{2}, Y_{GIH} \le y \le Y_{GIH} + Ly_{GIH} \right).$$

Here, $T_{F\ GIH}$ is the temperature of the side surfaces of the GIH. Thermocouple measurements show that the $T_{F\ GIH}$ value does not practically depend on the experimental conditions. By the 20th minute of the GIH operation, it reaches the value $T_{F\ GIH} \cong 47 \pm 4\ ^\circ\mathrm{C}$.

On the upper surface of the GIH:

$$\vec{\nabla T}(\tau,x,y) = -\frac{q_{F\ GIH}}{\lambda}, \tau > 0,$$
$$(x,y) \in \left(X_{GIH} - \frac{Lx_{GIH}}{2} \le x \le X_{GIH} + \frac{Lx_{GIH}}{2}, y = Y_{GIH} + Ly_{GIH} \right).$$

Here, $q_{F\ GIH}$—density of convective heat flux of combustion products. The q_{F_GIE} value was determined by the rated thermal power ($Q_{V\ GIH}$W), radiant efficiency (η_{Rad}) and the upper surface area of the GIH (F_{Up_GIH}, m^2) in accordance with the relationship: $q_{F\ GIH} = (1 - \eta_{\mathrm{Rad}}) Q_{V\_GIH} / F_{Up\_GIH}$.

The density of the heat flux to the surface q_{sol} was the sum of the density of the conductive–convective heat flux to this surface q_{gas} and the density of the radiative thermal q_{rad} from all radiating surfaces:

$$q_{sol} = q_{gas} + q_{rad}, \tau > 0,\ (x,y) \in A_{\Sigma\ F\ in} \cup A_{\mathrm{Tb}\ F}.$$

The no-slip conditions for the "gas-solid surface" interfaces were taken as the boundary conditions for the system of Equations (2) and (3) [22–24].

$$\vec{u}(\tau, x, y) = 0, \tau > 0,$$

$$(x,y) \in A_{\Sigma\ F\ in} \cup A_{\mathrm{Tb}\ F} \cup A_{\mathrm{GIH}\ F} \cup A_{\mathrm{VENT}\ F} \backslash A_{\mathrm{VENT}\ F\ out} \backslash A_{\mathrm{VENT}\ F\ in}.$$

Since the viscous effects prevail over turbulent ones near solid surfaces, the method of near-wall functions was used [22–24].

For the air inlet area $(x,y) \in A_{\mathrm{VENT}\ F\ out}$, the air velocity normal to the surface was set, which can be determined by the mass flow rate, the duct area and the inflow air temperature. For the air outlet area $(x,y) \in A_{\mathrm{VENT}\ F\ out}$, the pressure value was set equal to the atmospheric pressure outside the premises.

It was assumed that air pollution with carbon dioxide occurred due to the combustible gas burning and localizing on the upper GIH surface:

$$j_{CO_2}(\tau, x, y) = j_{GIHCO_2}, \tau > 0,$$

$$(x,y) \in \left(X_{GIH} - \frac{Lx_{GIH}}{2} \le x \le X_{GIH} + \frac{Lx_{GIH}}{2}, y = Y_{GIH} + Ly_{GIH} \right).$$

Based on the heat release rate from the chemical reaction $Q_{V\ GIH}$ and the net calorific value of the fuel Q_{fuel} (J/m^3), the volumetric flow rate of combustible gas for the initial temperature T_{beg} was determined as $V'_{fuel} = \left(Q_{V\ GIH}/Q_{fuel}\right)\left(T_{beg}/273\right)$. The standard fuel formula $C_{\gamma C}H_{\gamma H}$ was used to calculate its molar mass $M_{fuel} = \gamma C{\cdot}12 + \gamma H{\cdot}1$ and mass fraction of carbon $g_{C\ fuel} = \gamma C{\cdot}12/M_{fuel}$. The density of the pollution mass flow was calculated using the formula:

$$j_{GIH\ CO_2} = V'_{fuel} \cdot \frac{p_{beg} \cdot M_{fuel}}{8314 \cdot T_{beg}} \cdot g_{C\ fuel} \cdot \left(\frac{44}{12}\right) / F_{Up\_GIH}. \quad (11)$$

It was assumed that the air coming from the air-exchange system into the premises contains CO_2 in an amount equal to the initial value ω_{CO20}.

The no-flow condition was set for all other solid surfaces:

$$\vec{n} \cdot j_{CO_2}(\tau, x, y) = 0, \tau > 0,$$

$$(x,y) \in A_{\Sigma\ F\ in} \cup A_{\mathrm{Tb}\ F} \cup A_{\mathrm{GIH}\ F} \cup A_{\mathrm{VENT}\ F} \backslash A_{\mathrm{VENT}\ F\ out} \backslash A_{\mathrm{VENT}\ F\ in},$$

$$(x,y) \notin \left(X_{GIH} - \frac{Lx_{GIH}}{2} \le x \le X_{GIH} + \frac{Lx_{GIH}}{2}, y = Y_{GIH} + Ly_{GIH} \right),$$

where $\vec{n}$—unit normal vector to the surface.

The finite element method was chosen for solving the system of Equations (1)–(10) with the corresponding initial and boundary conditions. The COMSOL Multiphysics software environment with modules «The Heat Transfer in Fluids», «The Turbulent Flow» , «Transport of Concentrated Species» and «Surface-to-Surface Radiation» was used.

For a comparative analysis of changes in the average pollution of a room in mass (g_{CO_2}) and volume (r_{CO_2}) fractions, it was assumed that:

1. Mass flows G_{vent} of the air-exchange system (inlet and outlet) were the same;
2. The air mass on the premises m_{room} was constant;
3. The mass carbon dioxide influx on the premises due to combustion did not change the air mass on the premises. It was considered constant and determined by the following: $G_{GIH\ CO_2} = j_{GIH\ CO_2} \cdot F_{Up\_GIH}$.

Under these assumptions, the change in the CO_2 mass on the premises can be described by the following relationship:

$$m_{room}\frac{dg_{CO_2}}{d\tau} = -G_{vent}\left(g_{CO_2} - g_{0CO_2}\right) + G_{GIH\ CO_2} \quad (12)$$

Solutions were obtained by the following transformations:

$$\frac{dg_{CO_2}}{d\tau} = -\frac{G_{vent}}{m_{room}}\left[\left(g_{CO_2} - g_{0CO_2}\right) - \frac{G_{GIH\ CO_2}}{G_{vent}}\right],$$

$$\theta = \left(g_{CO_2} - g_{0CO_2}\right) - \frac{G_{GIH\ CO_2}}{G_{vent}},$$

$$\frac{dg_{CO_2}}{\theta} = -\frac{G_{vent}}{m_{room}}d\tau \Longrightarrow \theta = \theta_0 e^{-\frac{G_{vent}}{m_{room}}\tau},\ \theta_0 = -\frac{G_{GIH\ CO_2}}{G_{vent}}.$$

The $\frac{G_{vent}}{m_{room}}$ value has a clear physical meaning: $Kr = \frac{G_{vent}}{m_{room}} = \frac{\text{Air exchange rate}}{3600}$.
The solution for g_{CO_2} can be written as:

$$g_{CO_2} = g_{0\ CO_2} + \frac{G_{GIH\ CO_2}}{G_{vent}}\cdot\left(1 - e^{-Kr\cdot\tau}\right). \quad (13)$$

A steady-state value of $g_{CO_2\ max}$ is reached at $\tau \longrightarrow \infty$:

$$g_{CO_2\ max} = g_{0\ CO_2} + \frac{G_{GIH\ CO_2}}{G_{vent}}.$$

To determine r_{CO_2}, the following relationship was used:

$$r_{CO_2} = \frac{g_{CO_2}}{44}\cdot\frac{1}{\frac{g_{CO_2}}{44} + \frac{\left(1-g_{CO_2}\right)}{29}}. \quad (14)$$

The verification of the mathematical model of heat and mass transfer processes in a room with a gas infrared heater (GIH) was conducted using previously obtained experimental data. The experiments examined the thermal state and distribution of combustion products of natural gas (propane) used in a gas infrared heater. The temperatures and concentrations of CO_2, as well as the flow structure in local working areas of a large production facility, were observed.

Experimental studies were carried out in a room with overall dimensions of 5 × 4.4 × 11 m (Figure 2) with an installed light-type GIH-5 of 5 kW nominal thermal power (manufactured by the Sibshvank company, radiant efficiency η_{Rad} = 0.57). An experimental frame made of aluminum pipes with a diameter of 0.015 m with a plastic outer covering was installed in the room. Such a covering allowed us to place a horizontal wood panel (1.2 × 0.6 × 0.04 m) at different heights from the floor. The panel served as a model of the equipment. The highly thermally conductive material and small diameter of the tubes made it possible to neglect its effect on the generated thermal conditions in the room. The initial temperature was set equal to 7 °C.

The computer, shut-off, and control equipment, as well as a gas cylinder, were located outside the room to eliminate their influence on the thermal conditions of the temperature recording area. The used supply and exhaust ventilation system had an airflow of 420 m^3/h, an air speed of 2.3 m/s at the outlet of the supply channel, and an air temperature of 7 °C. The air ducts were located at a distance of 4 m from the floor (1 m above the GIH). The supply air temperature was kept constant by the operation of the electric air heater. Twelve chromel-alumel-type thermocouples with protection from re-radiation and an insulating coating of PFA fluoropolymer (junction thickness 0.08 mm) were used for temperature measurements. These thermocouples were placed in the local working area at various points in the zone of the GIH influence. The absolute error in temperature measurement was ±0.2 °C. Changes in CO_2 concentrations in the room and the local working area were recorded by three infrared (NDIR) CO_2 concentration-measurement sensors with a measurement range from 400 to 5000 ppm and an absolute measurement error of ±50 ppm. All the experiments under fixed conditions were carried out at least three times to ensure the possibility of estimating random errors. The standard deviations and corresponding coefficients of variation were calculated. The values of the latter for all experiments did not exceed 4%.

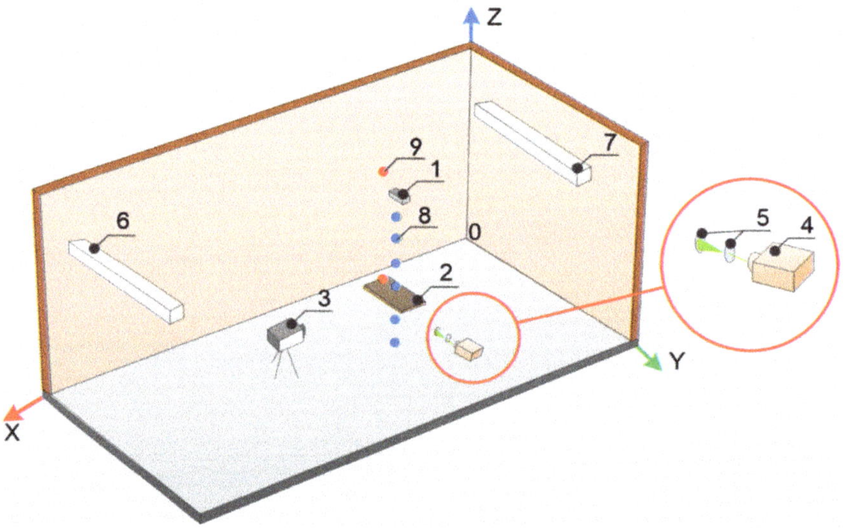

Figure 2. Scheme of the experimental box: 1—GIH; 2—horizontal panel (equipment model); 3—high-speed video camera; 4—laser; 5—system of mirrors for converting the laser beam; 6, 7—openings of the supply and exhaust ventilation system channels; 8, 9—sensors of temperature and CO_2 concentrations, respectively.

Figure 3 presents the main results obtained during the experimental studies.

The experiments have shown that when the surface of the panel is heated, the speed of the air above it increases and the main movement of air masses is directed upward (Figure 3a,b). This is because an increase in the temperature of the panel surface and, accordingly, the air around it (Figure 3c), leads to an increase in thermogravitational forces and air movement velocity and determines its direction (Figure 3a,b). The CO_2 concentration also increases both in the upper part of the room, where the combustion products of natural gas rise, and in the local working area (Figure 3c). This effect is most likely associated with the formation of a large-scale vortex in the room, which is formed during the joint operation of the GIH and the supply and exhaust ventilation system.

Figure 4 shows the distribution of temperatures at the height of the room on the axis of symmetry of the GIH influence zone and CO_2 concentrations in the local working area, established experimentally and numerically.

The deviation of the values of temperatures and CO_2 concentrations obtained during the modeling and experiments (Figure 4) does not exceed 10%. Therefore, it can be concluded that the created mathematical model is applicable for the further research and assessment of the influence of various parameters of the air-exchange system on the structure of air movement and distribution of heat and carbon dioxide in the room.

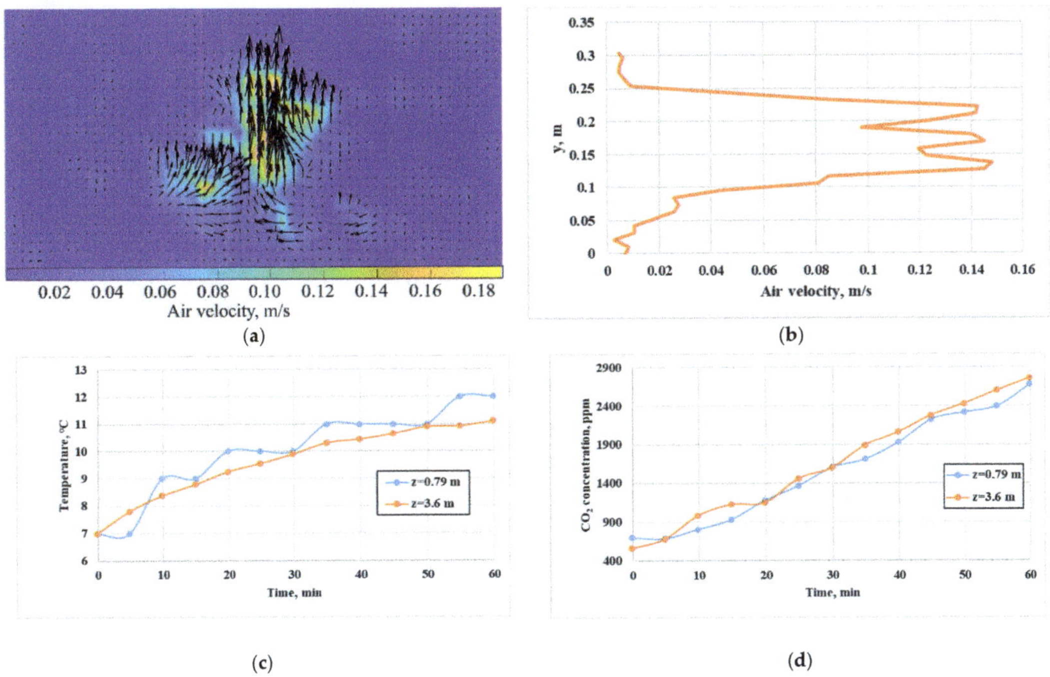

Figure 3. Fields of air movement vectors and speeds (**a**) above the surface of the equipment, speed profile (**b**) at the height from the horizontal surface of the equipment in the central section, as well as changes in air temperature (**c**) and CO_2 concentration (**d**) in time at points z = 0.79 m (local working area) and 3.6 m (upper area in the room at a distance of 0.8 m from the ceiling), y = 6.71 m; x = 3.5 m after 60 min of GIH operation.

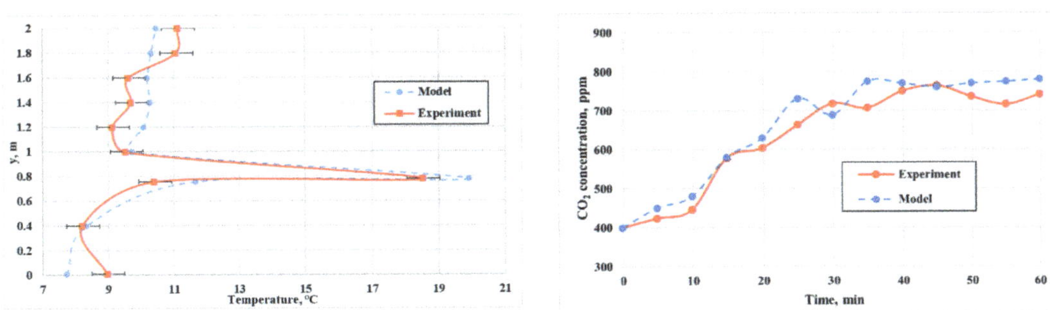

Figure 4. Distributions of experimental and theoretical values of air temperatures along the Y coordinate on the axis of symmetry GIH influence zone (**a**) and CO_2 concentrations in the local working area (z = 0.79 m, y = 6.71 m; x = 3.5 m) after 60 min of GIH operation (**b**).

3. Results

A premises with dimensions of 5 × 4.4 × 11 m was selected for modelling. This premises corresponded to the real one, where experimental studies had been previously carried out [28]. Within the framework of the two-dimensional approximation, a rectangular area with dimensions $L_x = 5 \times L_y = 4.4$ m, bounded by the floor, walls, and ceiling (enclosing structures, Table 2) was considered. The wall thickness was $L_{wall} = 0.1$ m. Two horizontal structural elements (Figure 1) corresponded to the GIH (dimensions Lx_{GIE} = 0.4 m, Ly_{GIE} = 0.05 m) and the equipment element—a horizontal panel (dimensions Lx_{tb} = 0.6 m, Ly_{tb} = 0.04 m).

Table 2. Thermophysical properties of building envelope materials and panels used in experiments [26,29].

Object	l, m	Material	ρ, kg m^{-3}	c, J kg^{-1} K^{-1}	λ, W K^{-1} m^{-1}	ε
Floor, ceiling and walls	0.1	concrete	2500	840	1.55	0.95
Horizontal panel	0.02	pine	520	2300	0.15	0.4

The gas infrared heater had a power of $Q_{VGIH} = 5$ kW. Propane (C_3H_8) with a calorific value $Q_{fuel} = 93,370$ kJ/m^3 was used as the working gas [26,29]. The CO_2 mass flow in this case was 3.1547×10^{-5} kg/s.

The value of 7 °C was taken as the initial temperature of all analyzed areas. The temperature of the incoming air from the air-exchange system was also 7 °C. The mass flow rate of the inlet part of the air-exchange system vent was varied in a range from 0.02 to 2 kg/s (9.09×10^{-5} to 9.09×10^{-3} kg/(s × m^3)).

The parameters from the COMSOL Multiphysics Materials Library were used as initial values for the air thermal properties. The main thermophysical parameters of the enclosing structures and the horizontal panel are presented in Table 2.

Previous studies showed that within 40–60 min from the start of a complex process of the mutual influence of thermogravitational and forced convection, caused by the GIH and air-exchange system operation, respectively, a stationary hydrodynamic picture of the air masses movement and the corresponding temperature field can be mainly established [1,28]. As the value of the average temperature over the volume of the premises is reached, the sufficiently intense changes in velocity and temperature ends. The average temperature over the volume corresponds to approximately 87% of the maximum value of the average temperature over the volume [28], which characterizes the stationary distribution of temperature fields. In this case, the representative fields of velocities and temperatures topologically and practically cease to change with time. Only the magnitudes of the maximum and minimum values change insignificantly. Therefore, the results of calculating the characteristics of processes by the time of 60 min are presented and analyzed below.

The fields of velocities, temperatures and mass concentrations of CO_2, as well as their distributions at the height of two vertical sections with coordinates x = 0.8 m (20 cm to the left of the panel) and x = 2.4 m (to the right of the panel 20 cm) in the possible working areas, are presented in Figures 5–14. In Figure 5, Figure 7, Figure 9, Figure 11 two white squares are the input and output areas of the air exchange system and the single rectangle is the radiator area (Figure 1). These areas are excluded from the simulation, and accordingly, they are white in the figures (there are no values for them, and in the color palette it is white color).

According to the standards, comfortable conditions are characterized not only by the temperature values but also by minimum temperature differences in height from the floor level to the upper limit of the local working area [10–13,30].

The flow pattern formed at the flow rate of blown air of 0.02 kg/s (9.09×10^{-5} kg/(s × m^3)) (Figure 5) is a consequence of air movement due to the air-exchange system operation and thermogravitational convection.

The results of numerical analysis (Figure 5) show that the cold air from the air inlet duct of the air-exchange system deviates from the horizontal path downwards. When flowing into the GIH, it is divided into two parts. Most of it deflects the upward flow of hot combustion products to the right towards the outlet duct. Some of these products, mixed with air from the air-exchange system and cooled near the right wall, move down.

Then, meeting with more heated surfaces of the floor and the horizontal panel surface, they form an upward flow. The second part of the air from the inlet part of the air-exchange system goes around the GIH from below. In the opposite direction, an upward flow moves from a heated horizontal surface of a panel that simulates the equipment. The interaction of these flows forms three main recirculation zones in the central area along the height of the premises. The currents flow clockwise on the left and right near the walls and counterclockwise in the central part. As a result, carbon dioxide accumulates in the ceiling area (mostly to the right of the GIH), then spreads, and, together with downward flows, falls down. At the same time, the asymmetry of the horizontal panel location and the ascending flows formed from it prevent the left flow with CO_2 from reaching the floor. On the right, the flow of CO_2 reaches the floor and gradually increases the degree of air pollution in the lower right corner of the region. Figure 6 shows the temperature and CO_2 concentration profiles in local work areas to the left and right of the equipment model.

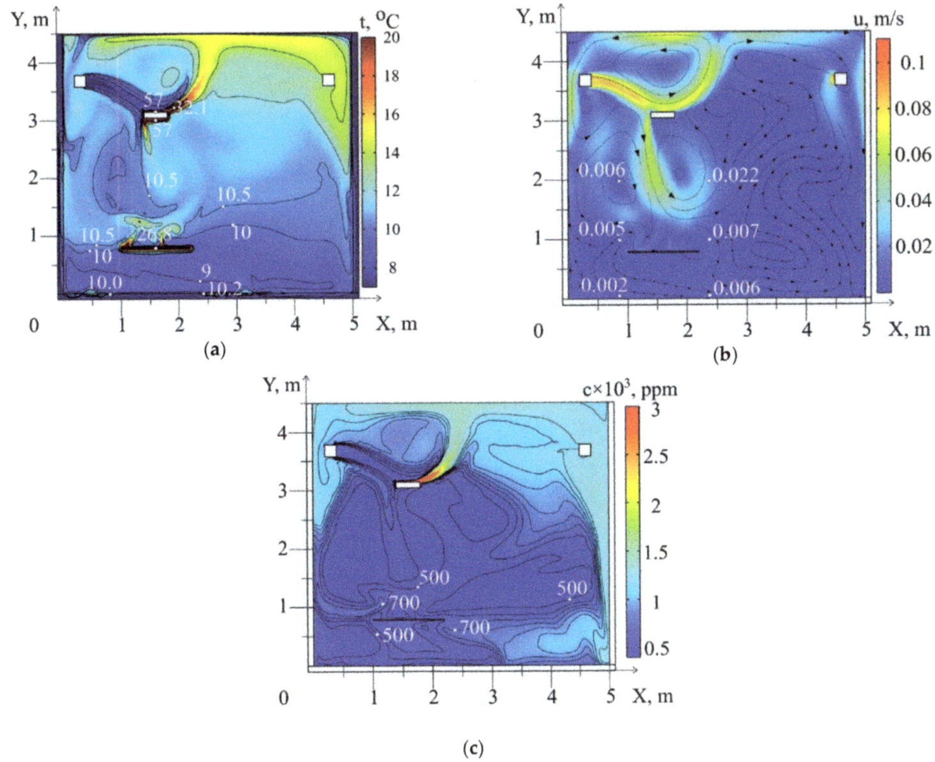

Figure 5. Fields of temperatures (**a**), air velocities (**b**) and CO_2 concentrations (**c**), formed by the 60th minute of the GIH operation at an airflow rate of 0.02 kg/s (9.09×10^{-5} kg/(s × m^3)).

An increase in airflow up to 0.04 kg/s (18.2×10^{-5} kg/(s × m^3)) changes the ratio between the effect of forced and thermogravitational convection (Figure 7). In this case, the air moves almost horizontally from the inlet to the outlet duct, and the air ascending from the horizontal panel reaches the air inlet duct of the air-exchange system. Two recirculation zones are formed in the area central to the height of the premises. Their formation is mainly influenced by the descending flows of air cooled by the walls and the ascending warm flow from the horizontal panel. Such a flow distribution increases the temperature and concentration of CO_2 in the central and left areas of the premises. As a result, it affects the distribution of the temperatures and concentrations of CO_2 at the height in the working area to the left and right of the horizontal panel (Figure 8).

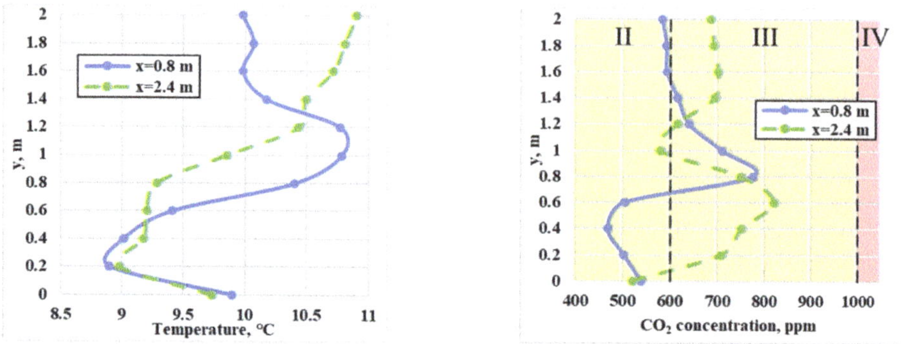

(a) (b)

Figure 6. Temperature profiles (**a**) and CO_2 concentration (**b**) formed by the 60th minute of the GIH operation at an airflow rate of 0.02 kg/s (9.09×10^{-5} kg/(s × m^3)). II–IV—indoor air quality class (Table 1).

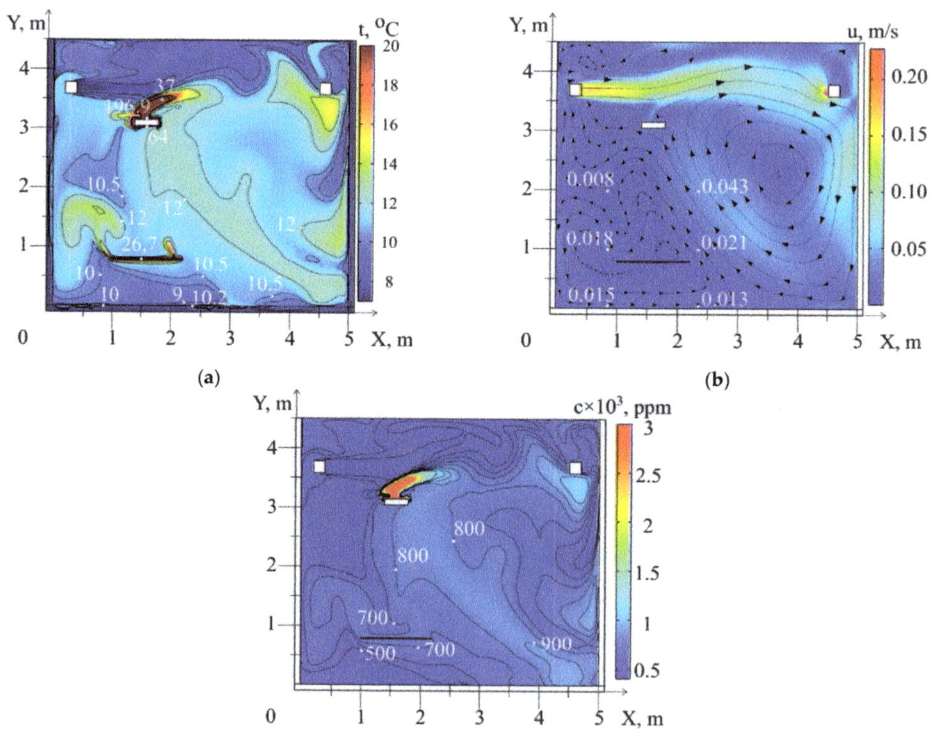

(c)

Figure 7. Fields of temperatures (**a**), air velocities (**b**) and CO_2 concentrations (**c**), formed by the 60th minute of the GIH operation at an airflow rate of 0.04 kg/s (18.2×10^{-5} kg/(s × m^3)).

At the same time, the air temperature in the local area to the right of the table rises to 14 °C (3 degrees (27%), compared with G_{vent} = 0.02 kg/s (9.09 × 10^{-5} kg/(s × m^3)). The CO_2 concentration also increases up to 870 ppm but does not exceed the acceptable value.

A further increase in the flow rate of the injected air up to 0.07 kg/s (31.8 × 10^{-5} kg/(s × m^3)) contributes to a significant increase in the recirculation flow intensity (Figure 9). This flow accumulates warm air in the central area of the premises and leads to an increase in temperature and carbon dioxide concentration in this part of the premises.

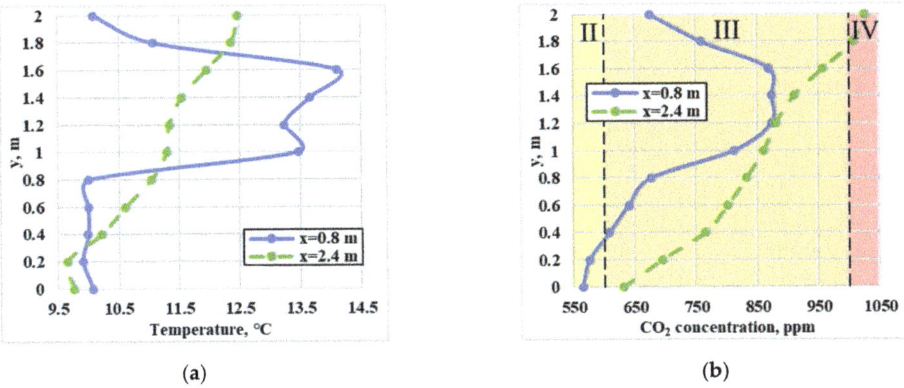

Figure 8. Temperature profiles (**a**) and CO_2 concentration (**b**) formed by the 60th minute of the GIH operation at an airflow rate of 0.04 kg/s (13.6 × 10^{-5} kg/(s × m^3)). II–IV—indoor air quality class (Table 1).

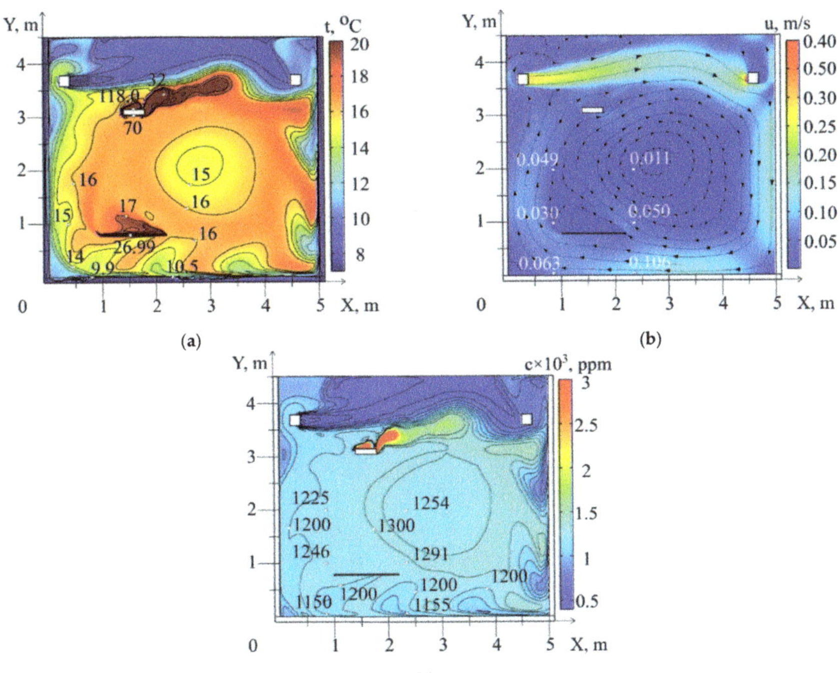

Figure 9. Fields of temperatures (**a**), air velocities (**b**) and CO_2 concentrations (**c**), formed by the 60th minute of the GIH operation at an airflow rate of 0.07 kg/s (31.8 × 10^{-5} kg/(s × m^3)).

An analysis of the obtained results shows that a significant increase in the arrival of a cold flow forms a pronounced recirculation flow along the enclosed structures, which forms relatively narrow near-wall areas near the latter. Warm air from the GIH and the horizontal panel accumulates in the peripheral part of the recirculation flow and gradually fills its central part with an increase in the discharge flow rate. It should be noted that, despite a sufficiently significant increase in air temperatures, air pollution in the working areas remains below the permissible values.

The results of the numerical simulation also show that the distribution of air heated from the GIH surface to the upper enclosing structure is hindered by a more intense flow of cold air from the air-exchange system, which moves almost horizontally to the middle of the study area.

In the section X > 3 m, the velocity of the airflow blown into the premises slows down (Figure 9b) to 0.1 m/s. The colder air from the inlet of the air-exchange system pushes down the heated layers of air (Figure 9a). As a result, warm air moves towards the right building envelope below the outlet of the air-exchange system.

As a result, with an increase in airflow up to 0.07 kg/s (31.8×10^{-5} kg/(s $\times$ m^3)), the air temperature in the central part of the premises rises, although the injected air has a significantly lower temperature than that in the working areas (Figure 10). But as a consequence of such a circulation flow, CO_2 also enters the local working area along with the air heated from the GIH. The gas concentration increases to the right of the panel from 850 ppm to 1200 ppm (higher than the allowable one) (Figures 8b and 10b).

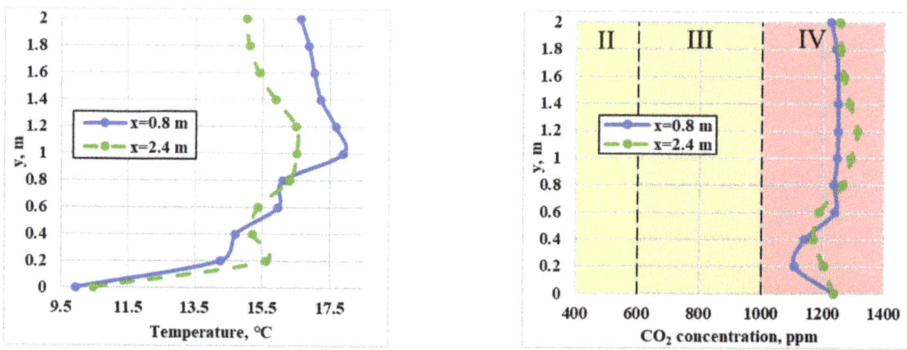

(**a**) (**b**)

Figure 10. Temperature profiles (**a**) and CO_2 concentration (**b**) formed by the 60th minute of the GIH operation at an airflow rate of 0.07 kg/s ($31.8 \cdot 10^{-5}$ kg/(s$\cdot$m^3)). II–IV—indoor air quality class (Table 1).

The subsequent increase in the air circulation ratio (40.9×10^{-5} kg/(s $\times$ m^3)) leads to the division of the solution area into two main circulation zones (Figure 11), located almost horizontally relative to each other.

The first zone is formed above the GIH. The blown air from the air-exchange system flows at a sufficiently high speed to the outlet area, practically without interacting with the air in the central part of the room. The second circulation zone is located below the inlet and outlet ventilation holes (Figure 11b). The air heated from the GIH case circulates in the lower part of the premises. The distribution of temperatures (Figure 12a) and concentrations (Figure 12b) in the local working area increases.

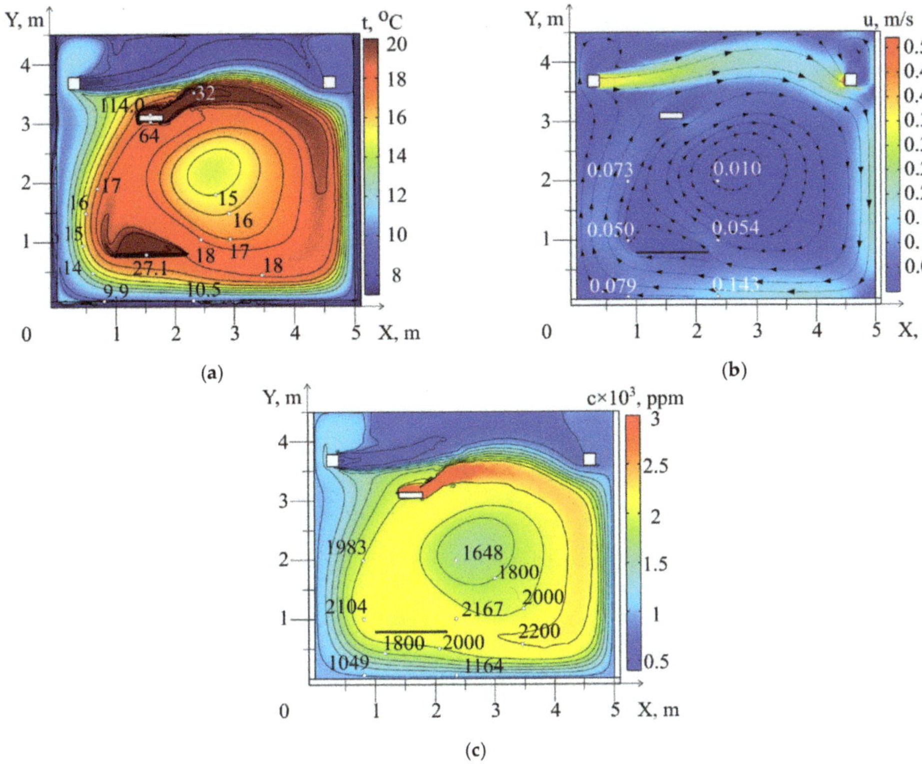

Figure 11. Fields of temperatures (**a**), air velocities (**b**) and CO_2 concentration (**c**), formed by the 60th minute of the GIH operation at an airflow rate of 0.09 kg/s (40.9×10^{-5} kg/(s × m^3)).

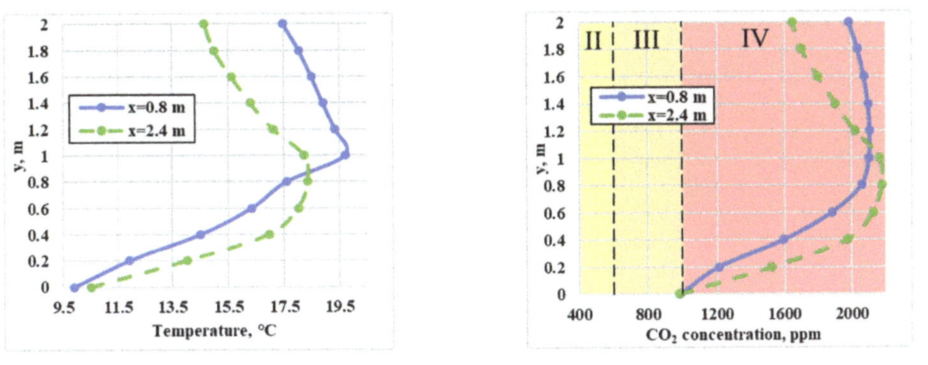

Figure 12. Temperature profiles (**a**) and CO_2 concentration (**b**) formed by the 60th minute of the GIH operation at an airflow rate of 0.09 kg/s (40.9×10^{-5} kg/(s × m^3)). II–IV—indoor air quality class (Table 1).

Figure 13 shows changes in the CO_2 average concentration in the premises volume, depending on the flow rate of the air-exchange system. The results were obtained on the basis of solving balance Equations (13) and (14) (used in the calculation methods of ventilation systems [5–8]).

An analysis of the results showed that for fairly typical premises with a working GIH and an air-exchange system, the standard concentrations for CO_2 content in the air (no more than 1000 ppm) are achieved at an airflow rate in the air-exchange system of at least 0.31 kg/s (Figure 13a). However, the excess of the limiting concentration of CO_2 in the premises volume is reached after 20–30 min at lower airflow rates.

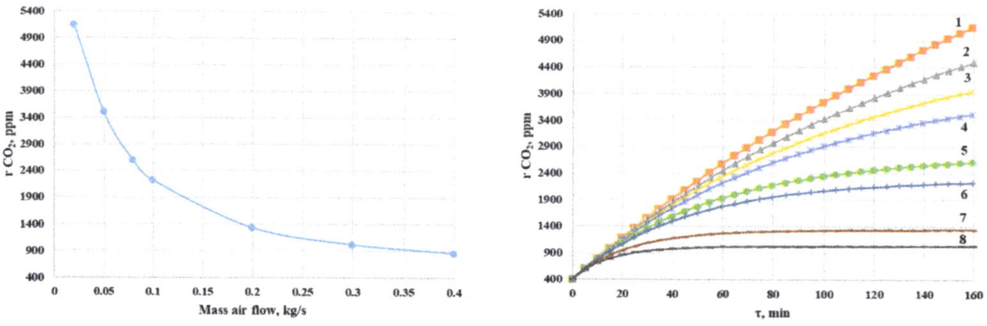

(**a**) (**b**)

Figure 13. Dependence of the average CO_2 concentration over the volume of the premises on the airflow in the air-exchange system (**a**) at τ = 160 min and its change in time (**b**) at airflow rates in the air-exchange system: 1—0.02 kg/s, 2—0.03 kg/s c, 3—0.04 kg/s, 4—0.05 kg/s, 5—0.08 kg/s, 6—0.1 kg/s, 7—0.2 kg/s, 8—0.3 kg/s.

At the same time, the simulation results (Figures 5–11) show that intensive air mixing in the room does not occur during the joint operation of the gas infrared heater and the air-exchange system (the main assumption in calculations using balance models). Zones with significantly different CO_2 concentrations are formed. Heated and carbon dioxide-polluted air is mainly located in the upper part of the premises. The CO_2 concentration values remain within acceptable limits in the local working areas. It can be concluded that when forming the thermal conditions in the local working area using a GIH, it is possible to maintain the standard conditions for CO_2 concentration with a lower consumption of air from the air-exchange system. This is especially true if the formation of a thermal area is required for a short time (from 30 min to 1.5 h). For example, according to the results of numerical simulation (Figure 14), with an airflow rate of 0.04 kg/s, a worker can stay in the local working area for quite a long time (about 120 min). However, according to balance models calculations, this time is 20 min and below (six times less) (Figure 13b).

A significant increase in airflow in the air-exchange system leads to an intensification of the air mass movement (mixing). In this case, both the temperature and the CO_2 concentration in the local working areas increase. For the range of airflow rates from 0.05 to 0.6 kg/s, CO_2 concentrations are above the critical level (from 1000 to 2000 ppm) (Figure 12). With a further increase in airflow in the air-exchange system to 0.1 kg/s, the temperature and CO_2 concentration in the local working area increase to the maximum (17 °C, 2000 ppm) (Figure 15). In the central part of the premises, an extensive circulation vortex is formed, into which combustion products from the GIH enter, but a further increase in the consumption of the air-exchange system leads to a decrease in both temperature (to 11–13 °C) and CO_2 concentration (tends to the initial—400 ppm) in local working areas (Figure 12b). This circumstance is a consequence of the increasing excess of the mass of incoming air over the flow of hot combustion products coming from the gas infrared heater.

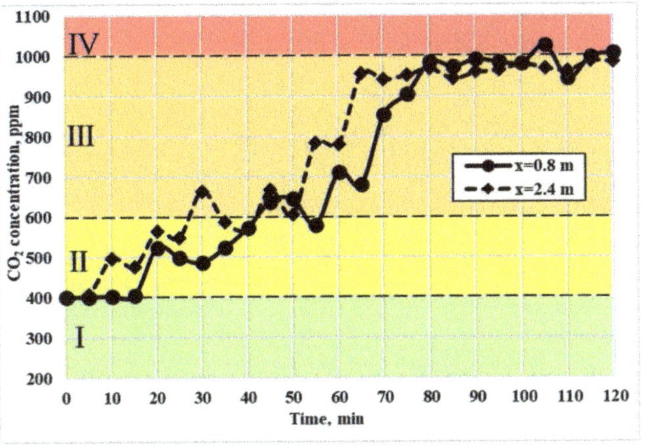

Figure 14. Dependence on time of the average value of CO_2 concentration at heights from 0 to 2 m in the working areas with a flow rate of ventilated air of 0.04 kg/s (18.2×10^{-5} kg/(s × m^3)). I–IV—indoor air quality class (Table 1).

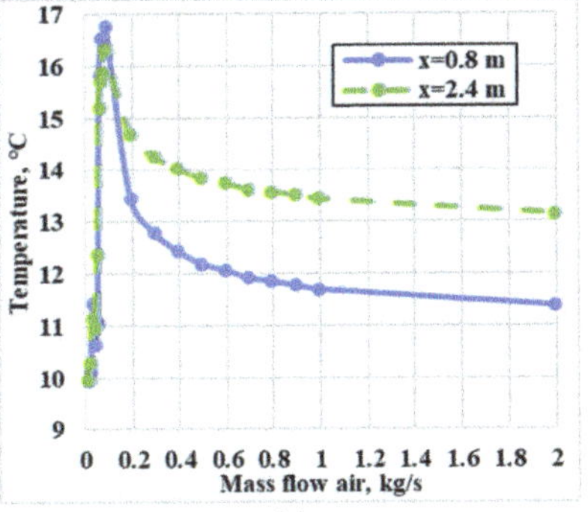

(**a**)

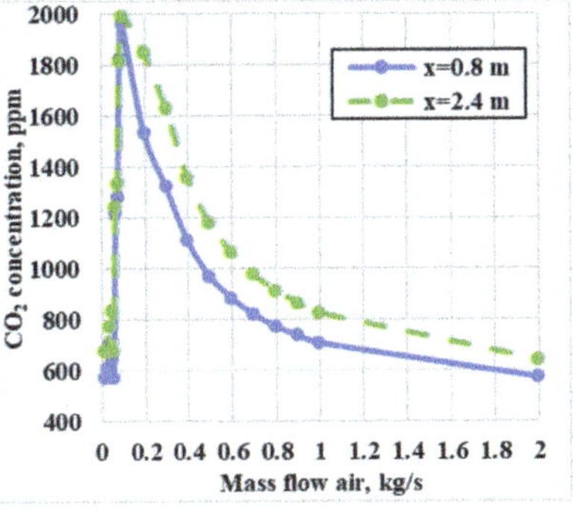

(**b**)

Figure 15. Dependence of the height-average air temperature (**a**) and CO_2 concentration (**b**) in local working areas to the left (x = 0.8 m) and to the right (x = 2.4 m) of the table by 20 cm from the airflow after 60 min of GIH operation.

The analysis of the numerical simulation results also showed that the necessary airflow rate from the air-exchange system to create regulated conditions for CO_2 concentration in the air (less than 1000 ppm) lies in the ranges from 0.02 to 0.04 kg/s and from 0.6 kg/s (Figure 15). The calculated air consumption according to the method based on balance models is 0.31 kg/s (outside this range).

4. Conclusions

A theoretical analysis of the effect of air-exchange system operation on the thermal conditions and carbon dioxide concentration of a local working area when using a light-type GIH was carried out. The numerical study of the processes was carried out using the COMSOL Multiphysics environment. Using the k-ε model made it possible to take into account the influence of turbulent flow. Changing the turbulence model from within the simulation environment range and its intensity parameters did not significantly affect the temperature and concentration fields in a larger volume of the room. This is due to the fact that the turbulence generation was localized in extremely limited volumes, positioned in the zone of forced air inflow from the air-exchange system and above the GIH. Such a case can be well described by the k-ε turbulence model. It was established that two zones with quite significantly different CO_2 concentrations and air temperatures are formed during GIH operation. The airflow in the air-exchange system significantly affects the formation of such zones in the room. At low airflow rates up to 0.04 kg/s (18.2×10^{-5} kg/(s × m^3)), a zone of air heated and highly concentrated by carbon dioxide is formed in the upper part of the room. It was found that with an airflow rate of 0.04 kg/s (13.6×10^{-5} kg/(s × m^3)), it is possible to direct some of the heated air to the local working area, increasing the air temperature by 27% without the CO_2 concentration exceeding the regulated value (less than 1000 ppm).

Author Contributions: Conceptualization, G.V.K.; methodology, V.I.M. and G.V.K.; software, V.I.M. and B.V.B.; validation, B.V.B., V.I.M. and F.Y.S.; formal analysis, V.I.M.; investigation, F.Y.S. and T.A.N.; writing—original draft preparation, F.Y.S.; writing—review and editing, T.A.N.; visualization, F.Y.S. and T.A.N.; supervision, G.V.K.; project administration, G.V.K.; funding acquisition, V.I.M. All authors have read and agreed to the published version of the manuscript.

Funding: This work is supported by the Russian Science Foundation (number grant 20-19-00226).

Data Availability Statement: Data available on request due to restrictions of institution.

Conflicts of Interest: The authors declare no conflicts of interest.

References

1. Kuznetsov, G.V.; Kurilenko, N.I.; Maksimov, V.I.; Nagornova, T.A. Experimental and numerical study of heat transfer in production area heated by gas infrared source. *Int. J. Therm. Sci.* **2020**, *154*, 106396. [CrossRef]
2. Dudkiewicz, E.; Szałański, P. Overview of exhaust gas heat recovery technologies for radiant heating systems in large halls. *Therm. Sci. Engin. Progress.* **2020**, *18*, 100522. [CrossRef]
3. Kavga, A.; Karanastasi, E.; Konstas, I.; Panidis, T. Performance of an Infrared Heating System in a Production Greenhouse. *IFAC Proc.* **2013**, *46*, 235–240. [CrossRef]
4. Dudkiewicz, E.; Fidorów-Kaprawy, N.; Szałański, P. Environmental benefits and energy savings from gas radiant heaters' flue-gas heat recovery. *Sustainability* **2022**, *14*, 8013. [CrossRef]
5. Mikhailova, L.Y.; Kurilenko, N.I.; Germanova, T.V.; Shcherbakova, E.N. Decentralized heat supply systems for industrial buildings using natural gas from the Arctic zone. *IOP Conf. Series. Earth Environ. Sci.* **2022**, *990*, 012066. [CrossRef]
6. Industrial Heating Systems by Schwank. Available online: https://schwank.co.uk/products/industrial-heating-systems/?lang=en (accessed on 15 November 2023).
7. Peng, P.; Zhang, C.; Li, W.; Pomianowski, M.; Gong, G.; Fang, X.; Chun, L.; Guo, R. Investigation on indoor airflow and contaminant dispersion of diffuse ceiling ventilation in heating and cooling modes. *J. Build. Eng.* **2023**, *80*, 107972. [CrossRef]
8. Zhang, C.; Pomianowski, M.; Heiselberg, P.K.; Yu, T. A review of integrated radiant heating/cooling with ventilation systems—Thermal comfort and indoor air quality. *Energ. Build.* **2020**, *223*, 110094. [CrossRef]
9. Maher, D.; Hana, A.; Arjmand, J.T.; Issakhov, A.; Sammouda, H.; Sheremet, M.; Sharma, S. Effect of inlet/outlet on thermal performance of naturally ventilated building. *Int. J. Low Carbon Technol.* **2021**, *16*, 1348–1362. [CrossRef]
10. *GOST 30494-2011*; Residential and Public Buildings. Microclimate Parameters for Indoor Enclosures: Moscow, Russia, 2011.
11. *EN 13779:2007*; Ventilation for Non-Residential Buildings. Performance Requirements for Ventilation and Room-Conditioning Systems: Moscow, Russia, 2007.
12. *ISO 7730*; Moderate Thermal Environment—Determination of the PMV and PPD Indices and Specification of the Conditions for Thermal Comfort. International Organization for Standardization: Geneva, Switzerland, 2005.
13. *EN 16798-1:2019*; Energy Performance of Buildings—Ventilation for Buildings—Part 1: Indoor Environmental Input Parameters for Design and Assessment of Energy Performance of Buildings Addressing Indoor Air Quality, Thermal Environment, Lighting and Acoustics—Module M1-6. CEN-CENELEC Management Centre: Brussels, Belgium, 2019.
14. Majumdar, D.; Chatterjee, S. Modelling accumulation of respiratory-CO_2 in closed rooms leading to decision-making on room occupancy. *MAPAN-J. Metrol. Soc. India* **2020**, *35*, 323–332. [CrossRef]
15. Zhang, D.; Zhang, L. Experimental and simulation research on indoor CO_2 removal efficiency and fresh air energy savings of living walls in office spaces. *Preprints* **2023**. [CrossRef]
16. Bivolarova, M.; Snaselova, T.; Markov, D.; Melikov, A.K. CO_2 based ventilation control—importance of sensor positioning, CO_2 Based Ventilation Control—Importance of Sensor Positioning. In Proceedings of the 15th ROOMVENT Virtual Conference, Torino, Italy, 15–17 February 2021.
17. Snaselova, T.; Bivolarova, M.; Markov, D.; Melikov, A. Exposure to metabolic CO_2 in a room with mixing air distribution. In Proceedings of the 15th ROOMVENT Virtual Conference, Torino, Italy, 15–17 February 2021.
18. Keli, A.; Melikov, A.K.; Bivolarova, M.P.; Mustakallio, P. Impact of room airflow interaction on metabolic CO_2 exposure. *E3S Web Conf.* **2019**, *111*, 02005. [CrossRef]

19. Borisov, B.V.; Vyatkin, A.V.; Kuznetsov, G.V.; Maksimov, V.I.; Nagornova, T.A. Mathematical modeling of heat transfer in a room with a gas infrared emitter, an air exchange system, and a local fence of the working area. *J. Appl. Ind. Math.* **2023**, *17*, 15–24. [CrossRef]
20. Borisov, B.V.; Vyatkin, A.V.; Kuznetsov, G.V.; Maksimov, V.I.; Nagornova, T.A. Numerical Analysis of the Influence of the Air Exchange System Configuration on the Temperature Regime of Local Working Areas in a Room with a Gas Infrared Heater. *Bull. Tomsk Polytech. Univ. Geo Assets Eng.* **2023**, *334*, 7–16. [CrossRef]
21. Bird, B.; Stewart, W.E.; Lightfoot, E.N. *Transport Phenomena*, 2nd ed.; John Wiley & Sons: Hoboken, NJ, USA, 2007.
22. Tritton, D.J. *Physical Fluid Dynamics*, 2nd ed.; Clarendon Press: Oxford, UK, 1988.
23. Wilcox, D.C. *Turbulence Modeling for CFD*, 2nd ed.; DCW Industries: Canada, CA, USA, 1988.
24. Kuzmin, D.; Mierka, O.; Turek, S. On the Implementation of the k-ε Turbulence Model in Incompressible Flow Solvers Based on a Finite Element Discretization. *Int. J. Comput. Sci. Math.* **2007**, *1*, 193–206. Available online: https://www.researchgate.net/publication/228529803 (accessed on 25 December 2023). [CrossRef]
25. Siegel, R.; Howell, J. *Thermal Radiation Heat Transfer*, 4th ed.; Taylor & Francis: New York, NY, USA, 2002.
26. Haynes, W.M. *Handbook of Chemistry and Physics 2015–2016*; CRC/Taylor & Francis: Boca Raton, FL, USA, 2015.
27. Curtiss, C.F.; Bird, R.B. Multicomponent Diffusion. *Ind. Chem. Res.* **1999**, *38*, 2515–2522. [CrossRef]
28. Borisov, B.V.; Vyatkin, A.V.; Kuznetsov, G.V.; Maksimov, V.I.; Nagornova, T.A. Analysis of the Influence of the Gas Infrared Heater and Equipment Element Relative Positions on Industrial Premises Thermal Conditions. *Energies* **2022**, *15*, 8749. [CrossRef]
29. DOE Fundamentals Handbook. *Thermodynamics, Heat Transfer and Fluid Flow (Volume 2 of 3)*; U.S. Department of Energy: Washington, DC, USA, 2016.
30. Karmann, C.; Schiavon, S.; Bauman, F. Thermal comfort in buildings using radiant vs. all-air systems: A critical literature review. *Build. Environ.* **2017**, *111*, 123–131. [CrossRef]

energies

Article

The Effects of Exterior Glazing on Human Thermal Comfort in Office Buildings

Bing Song [1,2], Lujian Bai [3,*] and Liu Yang [1,2]

1 College of Architecture, Xi'an University of Architecture and Technology, Xi'an 710055, China; songbing@xauat.edu.cn (B.S.); yangliu@xauat.edu.cn (L.Y.)
2 State Key Laboratory of Green Building, Xi'an University of Architecture and Technology, Xi'an 710055, China
3 College of Architecture and Civil Engineering, Xi'an University of Science and Technology, Xi'an 710054, China
* Correspondence: bailujian@xust.edu.cn

Abstract: As a major component of the building envelope, the energy-saving design of exterior windows is key to energy savings in office buildings. The conventional design of exterior windows mainly focused on their impact on heating and cooling energy but ignored the indoor thermal comfort problems caused by the direct solar radiation transmitted by windows and the fluctuation of their internal surface temperatures. This study analyzed the influence of exterior windows on the indoor thermal environment of office buildings by carrying out field experiments. The experiments were carried out in a typical office building in Xi'an during December and January. The impact of exterior windows on human thermal comfort was studied from two perspectives: longwave radiation from the surface of window glass and solar shortwave radiation. It was found that solar radiation was the main cause of temperature fluctuations on the internal surface of windows and created non-uniform thermal environments. The mean radiant temperature fluctuations in the near-window area could reach up to 7.8 °C due to outdoor solar radiation in winter. Solar radiation transmitted by windows directly affects thermal sensations. Since conventional thermal comfort models or indices underestimated the thermal sensations of occupants in the presence of solar radiation, the additional thermal loads caused by solar radiation needed to be taken into account. The allowable operative temperature range for maintaining thermal comfort should be reduced by 0.5 °C when occupants are exposed to solar radiation.

Keywords: thermal comfort; solar radiation; windows; office building

1. Introduction

With economic development and accelerated urbanization, building energy consumption in China has increased rapidly during the past few decades. According to the report published by THUBERC, the current energy consumption of buildings in China has reached 1.06 billion tons of standard coal, accounting for 21% of the country's total energy consumption. The energy consumption of office buildings, which accounts for 14% of the total building energy consumption, was one of the main reasons for the rapid increase in building energy consumption in recent years [1]. Thus, the energy-saving design of office buildings has become a matter of urgent concern. Unlike residential buildings, office buildings have a higher proportion of window area in the envelope. The impact of exterior windows on the energy consumption of office buildings is reflected in three aspects: illumination, heating, and air conditioning. In recent years, the exterior window area of office buildings has become larger, owing to various demands such as aesthetics, lighting, and psychology [2,3]. The increase in window area is beneficial for lighting energy consumption reduction, but it can lead to a significant increase in heating and air conditioning energy consumption [4,5].

The conventional energy-saving design for office building exterior windows mainly aims to reduce heat conduction by reducing the thermal conductivity of windows and to maintain the indoor air temperature within a comfortable range for the human body by reducing the energy consumption of heating and air conditioning [6–8]. However, approximately 50% of the body's heat is exchanged with indoor surroundings in the form of radiation, in which exterior windows play an important part and may affect indoor thermal comfort [9]. The conventional energy-saving design of exterior windows focuses on their impact on indoor air temperature but ignores the indoor thermal comfort problems caused by the solar radiation transmitted by windows and the fluctuation in their interior surface temperatures [10–12].

Owing to differences in the thermophysical properties of materials, the internal surface temperature of windows is prone to large fluctuations. According to a survey conducted by Ge and Fazio in Montreal, the interior surface temperature of exterior windows was only 10 °C and 3.8 °C when outdoor temperatures were −18 °C and −32 °C, respectively [13]. Changes in the interior surface temperature of windows can have an impact on the indoor radiant temperature and thermal comfort by longwave radiation. Gan found that the asymmetric radiation temperature 1 m from a window exceeded 10 °C when the outdoor air temperature was −4 °C and the indoor air temperature was 21 °C [14]. Chaiyapinunt's study of the indoor thermal environment of office buildings in Bangkok found that an increase in the interior surface temperature of windows can trigger an increase in the PPD value of the neighboring area, which in turn increases the risk of indoor overheating [15]. The numerical simulation analysis by Sengupta and Chapman found that only 7% of the room area in rooms with a window-to-wall ratio of 40% can meet the needs of thermal comfort in summer with solar radiation [16,17]. Too high or too low the surface temperature of the window can cause thermal discomfort and increase the operating time of air conditioners and other equipment, resulting in an increase in building energy consumption.

Shortwave radiation aroused by solar radiation through exterior windows is another important factor that triggers indoor thermal discomfort. The research by Zhao showed that the mean radiant temperature will increase by 4 °C when the solar radiation through exterior windows achieves 750 W/m^2 during the summer [18]. Chaiyapinunt indicated that the thermal discomfort caused by solar radiation through windows did not depend on the distance between the person and the window but on the magnitude and direction of solar radiation [19]. Arens found that when a person sitting near a window was exposed to direct solar radiation, the heat load acting on his or her body was equivalent to an increase in the average radiation temperature of the environment by 11 °C [20]. Through actual testing, Schutrum [21] figured out that on a cloudy day, the overall thermal sensation of the human body increased by 1.1 units when the window temperature increased from 3 °C to 48 °C. When sunlight directly struck the human body on a sunny day, the window temperature rose to 31.7 °C, and the overall thermal sensation of the human body increased by 2.5 units. The study by Lyons [22] further confirmed that the longwave radiation between a person and a window had the greatest impact on human thermal sensation when there was no sunlight exposure. However, when there was sunlight exposure, solar radiation became the dominant factor. Solar shortwave radiation transmission not only has an impact on the thermal comfort of the human body in the direct sunlight zone but also has a direct impact on the thermal uniformity of an entire room. Through numerical simulation analysis, Sengupta [17] found that exterior windows can lead to significant inhomogeneities in indoor thermal comfort distribution. Moreover, the heating or cooling system of an entire room cannot eliminate or improve the uneven distribution of thermal comfort. A study by Bessoudo and Tzempelikos et al. found that the internal surface temperatures of exterior windows rapidly increased and triggered asymmetric indoor radiation temperatures of up to 15 °C due to solar radiation [23,24]. Solar radiation transmitted by exterior windows will further exacerbate the inhomogeneity of the indoor thermal environment and indoor thermal discomfort and increase the intensity and duration of the use of indoor air conditioning systems, thus increasing the energy consumption of a building.

As a key component of the office building envelope, determining an appropriate thermal performance of windows, such as U-value and SHGC, would significantly reduce building energy consumption and improve indoor thermal comfort. Although the impact of exterior windows on indoor human thermal comfort has gradually attracted attention, the existing standards and specifications still use conventional human thermal comfort laws to guide indoor thermal environment design [25–27], which may limit the enhancement of energy-saving performance of exterior windows in office buildings and the improvement of indoor thermal environment. Established studies are mostly focused on theoretical analysis and lack on-site experimental research on occupants, particularly the thermal effect of solar radiation on occupants. Therefore, the aim of this study was to conduct on-site experiments to quantitatively analyze the impact of exterior windows on the indoor thermal environment of office buildings. The effect of exterior windows on human thermal comfort is studied from the perspectives of longwave radiation caused by the fluctuation in interior surface temperatures of windows and solar shortwave radiation caused by the direct solar radiation transmitted by windows to provide guidance and support for the energy-saving design of exterior windows in office buildings.

2. Method

To study the effect of exterior windows on the indoor thermal environment and thermal comfort, experiments were carried out in a typical office building during December and January. The experiments included two parts: (a) a questionnaire survey; and (b) a thermal environment test.

2.1. Experimental Site

The experimental site is located in a south-facing room of a typical office building in Xi'an, Shaanxi Province, China. Xi'an has a temperate climate influenced by the East Asian monsoon, classified under the building thermal climate zones as situated in the cold zone (CZ) [28]. The general climate of Xi'an is characterized by cold, dry winters, hot, humid summers, and dry springs and autumns. The most intense direct solar radiation occurs in winter as the dry air tends to overheat in rooms with heating systems and large exterior windows.

The experimental chamber plan is shown in Figure 1. The total area of the chamber was 63 m^2 with a height of 2.7 m. There are three large windows on the south wall with window areas of 5.46 m^2, 6.3 m^2, and 6.3 m^2. These windows consist of aluminum frames and double-layer glass. The window-to-wall ratio of the south wall was approximately 82%. The exterior windows were unobstructed outside. The north wall was a gypsum board partition wall, and a bookshelf on the west side of the room was backed by a gypsum partition wall. The lighting load of the chamber was 8.6 W/m^2, and half of the lighting was switched on during the experiment. The heating system was set at 20 °C during the experiments.

2.2. Questionnaire Survey

A questionnaire survey is an effective method for investigating the impact of the thermal environment on subjective thermal sensations. The questionnaire consisted of two parts: (a) basic information of personal factors, including participant's age, activity status, and clothing; and (b) subjective thermal sensation, thermal comfort. According to ASHRAE Standard-55 [29], the thermal sensation scale is a symmetrical and bipolar seven-point scale. The seven-point scale is perfectly sufficient for general indoor thermal environment studies. However, it may not be accurate for environments containing strong solar radiation heat sources. Therefore, a nine-point scale was used in the questionnaire design, as shown in Figure 2a [30]. The acceptable scale in the questionnaire to count occupants' satisfaction and dissatisfaction is shown in Figure 2b.

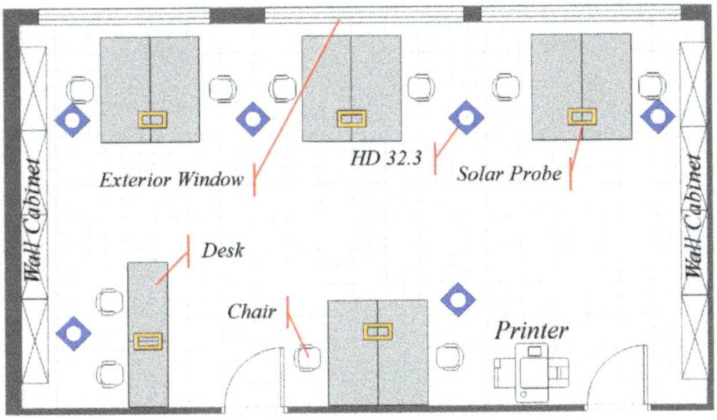

Figure 1. Experimental chamber plan and the arrangement of furniture and equipment.

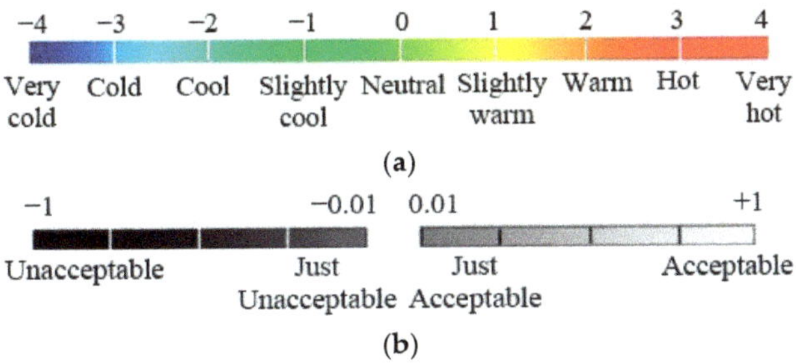

Figure 2. The thermal sensation scale and acceptable scale applied in the questionnaire survey. (**a**) Nine-point thermal sensation scale in the questionnaire. (**b**) Acceptable scale in the questionnaire.

Subjects reported their clothing levels at the time of completing the questionnaire by means of a clothing checklist in the questionnaire. The values of clothing insulation were quantified in units of clo based on the evaluation standard for the indoor thermal environment in civil buildings (GB/T 50785) [27]. The added insulation when sitting on the chair was estimated to be 0.1 clo according to the GB/T 50785.

2.3. Subject Characteristics

A total of 33 subjects participated in the whole experiment, including 16 males and 17 females. The BMI index of the subjects was mainly concentrated between 21 and 24, which is in line with the characteristics of the general population in an office environment. The general information on these subjects can be found in Table 1. The metabolic rate was determined by the activity type of the occupant and estimated to be 1.2 met in this study, which corresponds to office activities and seated according to GB/T 50785.

2.4. Measurement Points

The purpose of the test was to observe the direct solar radiation transmitted through the exterior windows and the variation in mean radiant temperature. The test was divided into two parts: indoor and outdoor. Indoor test items included air temperature, relative humidity, air velocity, solar radiation, and internal surface temperature of the envelope. Outdoor test items included total solar radiation, diffuse solar radiation, total solar radiation in south vertical planes, and air temperature. The arrangement of indoor equipment is shown in Figure 3. The height of equipment for air temperature, relative humidity, and air velocity was 0.6 m from the ground level. The main equipment used for testing is shown in Table 2.

Table 1. Personal factors of the subjects in the survey.

Items		Male	Female	Total
Clothes (clo)	Max	1.12	1.14	1.14
	Min	0.79	0.78	0.78
	Mean	0.94	0.96	0.95
	SD	0.11	0.13	0.12
Age	Max	38	36	38
	Min	23	21	21
	Mean	26.2	25.6	25.9
	SD	3.67	4.15	3.82
Metabolic rate (met)			1.2	

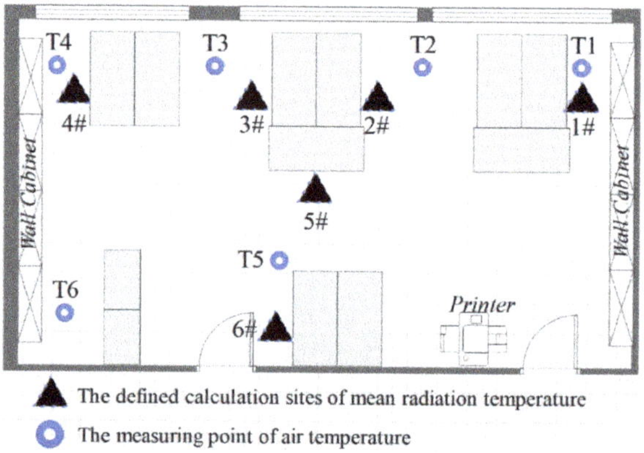

Figure 3. The defined calculation sites of mean radiant temperature and the measuring point of air temperature in the experimental chamber.

Table 2. Equipment used in on-site experimental research.

Instrument	Variables	Range and Accuracy
HD 32.3 (Delta OHM, Veneto, Italy)	Air temperature Relative humidity Air velocity	−40~100 °C (1/3 DIN) 0~100% (±1.5%) 0 ∼ 5 m/s (±0.2 m/s)
UX120-014M with the thermocouple wire of type K (Onset, Cape Cod, MA, America)	Surface temperature of envelope	−20~70 °C (±0.7 °C)
H21-USB (Onset, Cape Cod, MA, America)	Air temperature	−40~75 °C (1/3 DIN)
Delta LP PYRA (Delta OHM, Veneto, Italy)	Solar irradiance of outdoor and indoor environment	0~2000 W/m^2 spectrum from 305 nm to 2800 nm

2.5. Experimental Procedure

The seat for each subject was randomized during the experiment, and up to ten subjects could be accommodated in a single experiment due to chamber seating limitations. To eliminate the impact of previous thermal experiences, the subjects were asked to sit in an adjacent room with the same thermal environment as the experimental chamber for 30 min before the experiment to fully adapt to the current thermal environment. During the experiment, subjects remained seated and filled out the questionnaire at 30 min intervals.

2.6. Data Processing Method

2.6.1. Mean Radiant Temperature

Radiant heat transfer is the primary means by which external windows affect human thermal comfort. According to the radiation wavelength, the radiant heat transfer can be classified as longwave or shortwave. Longwave radiant heat transfer is the result of the temperature of the internal surface of an exterior window, while shortwave radiant heat transfer is the result of transmitted solar radiation.

In previous studies on human thermal comfort, the mean radiant temperature (T_{r-long}) has usually been used to measure the longwave radiant heat transfer of an entire enclosed space to the human body under the action of longwave radiation alone [31]. In this study, the mean radiant temperature was calculated from the test surface temperatures of each envelope according to the following equation:

$$\overline{T}_{r-long}^4 = T_1^4 F_{p-1} + T_2^4 F_{p-2} + \cdots + T_i^4 F_{s-i}, \quad (1)$$

The F_{s-i} can be calculated according to the method used in [31]. The effect of a temperature change on the internal surface of an exterior window on thermal comfort is mainly reflected in the change of T_i in Equation (1).

2.6.2. Solar-Adjusted Mean Radiant Temperature

The thermal effect of solar radiation on occupants is an additional heat load applied to the human body. In recent years, many scholars have proposed a variety of calculation models from different perspectives [32–37]. There are three main categories: (a) converting the transmitted solar radiation into the mean indoor radiant temperature increment [15,32–36], (b) directly substituting the transmitted solar radiation into the *PMV* calculation as the additional heat load on the human body [12,37], and (c) converting the transmitted solar radiation into an indoor air-temperature increment [38]. Based on the mechanism of the thermal effect of solar shortwave radiation on the human body, it is more reasonable to convert the transmitted solar radiation into a mean indoor radiant temperature increment. Therefore, this study calculated the thermal effect of solar shortwave radiation on the human body as an equivalent increase in the mean radiant temperature. The corrected mean radiant temperature T_{smrt} was written as [33]:

$$T_{\text{smrt}} = \sqrt[4]{\sum_{i=1}^{N} F_{s\to i} T_i^4 + \frac{1}{\varepsilon_S \sigma}\left(\alpha_{\text{irr}} \sum_{j=1}^{M} F_{s\to j} I_{\text{d}j} + \alpha_{\text{irr}} f_{\text{p}} I_{\text{b}}\right)} \quad (2)$$

The α_{irr} was taken as 0.67 in this study.

2.6.3. Data Analysis

The operative temperature is primarily used to measure the indoor thermal environmental conditions for accessing thermal comfort. The operative temperature in this study was calculated based on the following equation [31]:

$$T_o = \frac{h_c t_a + h_r \overline{T}_r}{h_c + h_r} \quad (3)$$

In this study, there were two values for T_r. When T_r was the mean radiant temperature, the calculated value was the operative temperature. When T_r was the solar-adjusted mean radiant temperature, the calculated value was the solar-adjusted operative temperature (T_{so}).

The Pearson correlation coefficient was used to test the degree of linear correlation between the two sets of data and was calculated as:

$$r = \frac{\sum_{i=1}^{n}(X_i - \overline{X})(Y_i - \overline{Y})}{\sqrt{\sum_{i=1}^{n}(X_i - \overline{X})^2}\sqrt{\sum_{i=1}^{n}(Y_i - \overline{Y})^2}}, \quad (4)$$

3. Result Analysis

3.1. Environmental Test Results

As glass facades have an effect on solar radiation, there may be some differences in air temperatures at different locations. Table 3 compares the air temperatures of each measuring point shown in Figure 3. It can be found that the magnitude of air temperature fluctuations decreased with the increasing distance from the windows. Measuring points near the windows had higher average temperatures than those away from the windows, although the difference was small. Therefore, the following operative temperatures and *PMV* values of each subject were calculated according to the data measured by the nearest measuring point.

Table 3. Summary of indoor air temperatures.

Variables	T1 (°C)	T2 (°C)	T3 (°C)	T4 (°C)	T5 (°C)	T6 (°C)
Max.	27.1	27.4	27.6	26.9	26.1	25.9
Min.	17.2	17.5	17.3	16.9	18.9	18.5
Mean	22.9	23.0	23.1	22.7	21.9	21.6
Median	22.3	22.6	22.4	22.1	22.0	21.7
SD	1.85	1.87	1.92	1.82	1.73	1.69

The indoor thermal environment test results for three consecutive days were analyzed, including three different outdoor working conditions: sunny, cloudy, and overcast. Day 1 was overcast, Day 2 was sunny, and Day 3 was cloudy. The analysis included the dynamic trends of the internal surface temperature of the window, the indoor air temperature, and the mean radiant temperature at different locations in the room. The results are shown in Figure 4, and the defined calculation sites of mean radiant temperature in the tested room are shown in Figure 3.

As shown in Figure 4a, the outdoor solar radiation had a significant disturbance effect on the indoor thermal environment. The fluctuation amplitude of the indoor thermal environment was directly related to the outdoor solar radiation intensity during the same period. The most significant impact of outdoor solar radiation was on the temperature of the interior surface of the window glass. As the intensity of outdoor solar radiation increased, the temperature of the interior surface of the window increased rapidly up to 34.6 °C and had large temperature fluctuation. The test results of Day 2 showed that the fluctuation amplitude of the surface temperature of the window glass could reach 25.2 °C between 8:00 and 19:00. Fluctuations in the surface temperature of the window glass had a direct impact on the indoor air temperature and mean radiant temperature. According to Section 2.6.1, an increase in the surface temperature of the window glass can directly trigger an increase in the mean indoor radiant temperature. Figure 4a shows that on a clear winter day, the impact of outdoor solar radiation caused the fluctuation amplitude of the mean indoor radiation temperature near the window to reach 7.8 °C. A comparison of the analysis results under different outdoor environments shows that the higher the outdoor solar radiation intensity, the more fluctuations in the indoor thermal environment.

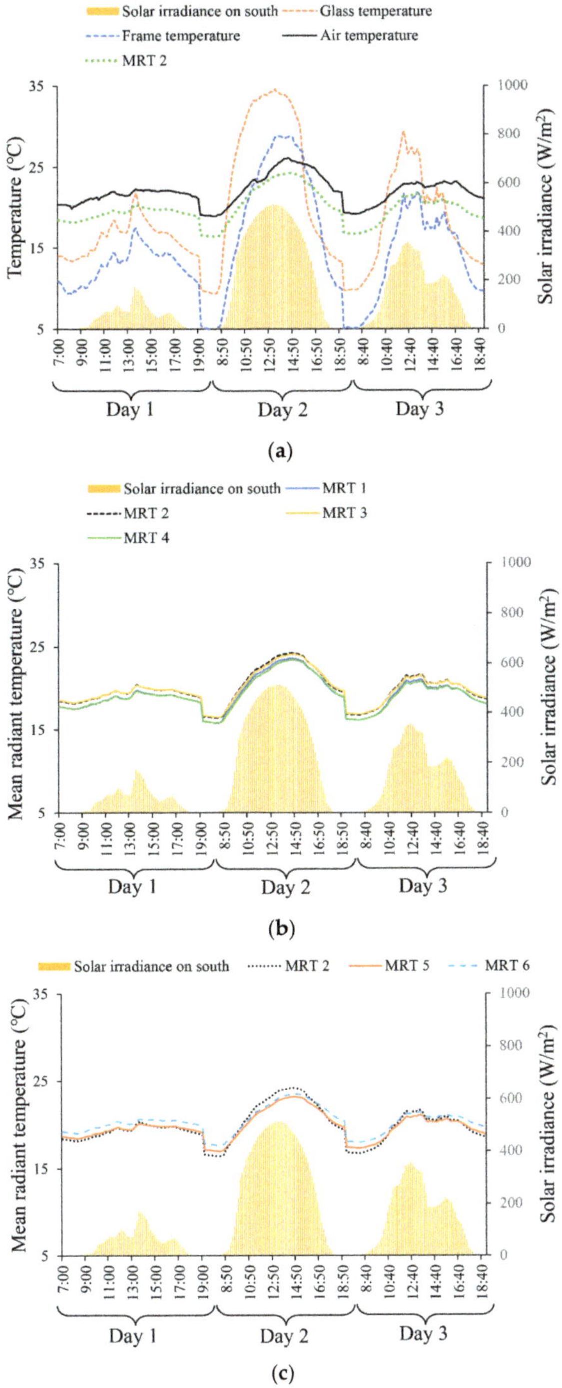

Figure 4. Indoor thermal environment test results. (**a**) Comprehensive analysis result. (**b**) Comparative analysis result of positions in the direction of room openings. (**c**) Comparative analysis of positions in the direction of room depth.

As demonstrated by the variation of the mean radiant temperature at different locations in the room in Figure 4b,c, a significant difference in the mean radiant temperature from the window existed at different locations. The maximum difference in average radiation temperature between Points 2 and 5 was 1.1 °C. Throughout the day, the fluctuation amplitude of the mean radiant temperature gradually weakened as the distance from the window increased. The fluctuation amplitudes of the mean radiant temperature at Points 2, 5, and 6 were 7.8 °C, 6.2 °C, and 5.8 °C, respectively. Figure 4b shows that at the same distance from the window, the mean radiant temperature in the central area of the room was more significantly affected by the surface temperature of the window than that in the areas on both sides. This was due to the larger visual coefficient between the central area and the window compared to other positions. This also indicates that for office buildings with large windows, the central area of a window is the most unfavorable area for the indoor thermal environment. As shown in Figure 4c, the mean radiant temperature generally increased with the distance from the window during Day 1. It was indicated that the glass surface temperature was mainly influenced by the outdoor air temperature on an overcast day. The lower glass temperature would lead to a lower mean radiant temperature in the vicinity of the window.

The analysis results of the indoor thermal environment suggest that the fluctuation of the window glass surface temperature was the main cause of changes in indoor mean radiant temperature. Additionally, this effect depended on the outdoor solar radiation intensity. The scope of the impact of windows on the indoor thermal environment was limited and concentrated in the near-window region. Reducing the heat-transfer coefficient of windows as well as the absorption and re-conduction of solar radiation by window glass was an effective way to weaken the impact of windows on indoor thermal comfort and to maintain indoor thermal stability. As introduced above, the conventional design of exterior windows mainly focused on their impact on heating and cooling energy. This study showed that the occupants' thermal comfort demands should also be considered in determining the optimum heat-transfer coefficient for exterior windows in the future. Considering the non-uniform indoor thermal environment caused by windows, office buildings with large window areas should be divided into zones for air temperature regulation, or personalized tools should be used to regulate the human thermal sensation near windows. Moreover, more and more new technologies and materials, such as smart electrochromic windows, thermochromic smart windows, and aerogel glazing systems, have been applied to building windows in recent years [39–41]. The impact of windows on indoor thermal comfort will be weakened by the excellent capability of solar management for these new window systems.

3.2. *Psychological Responses*

For the analysis of thermal sensation, the data obtained during the experiment were analyzed based on the presence or absence of direct sunlight. In Section 3.2.1, the general thermal sensation of occupants only under longwave radiation caused by temperature fluctuations on the window's inner surface is analyzed. Section 3.2.2 compares the trends in occupant thermal sensations when exposed to direct solar radiation under similar thermal environmental conditions.

3.2.1. Thermal Comfort without Direct Solar Radiation

As mentioned above, the variation of indoor air and ambient radiation temperatures of office buildings are the main impacts on indoor thermal comfort. The impact of changes in the indoor thermal environment on thermal sensation can be analyzed by establishing the variation trends between subjective thermal sensations and the operative temperatures over the same period. The distribution of subjective thermal sensation under each operative temperature is shown in Figure 5.

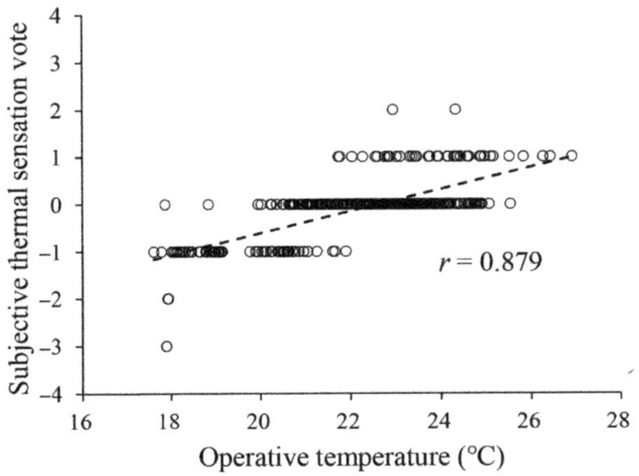

Figure 5. The relationship between subjective thermal sensation votes and operative temperatures.

As shown in Figure 5, with the increase in operative temperature, the subjective thermal sensation increased progressively. If the subjective thermal sensation vote (TSV) values of −1 and 1 were taken as the acceptability limits, the allowable operative temperature range is 18–25 °C. However, the upper limit of 25 °C is not credible owing to sample size limitations. If the comfort range was defined based on the neutral thermal sensation, the allowable operative temperatures should be within the range of 21–25 °C.

Although the thermal sensation vote can reflect the subjective thermal sensation in an indoor thermal environment, it cannot fully reflect occupants' satisfaction and comfort because of differences between individuals' thermal preferences. Previous studies have shown a good correlation between overall thermal sensation and thermal comfort in a uniform thermal environment, but a certain deviation exists between the two in non-uniform thermal environments [42]. Therefore, determining whether the indoor thermal environment induced by windows can meet the thermal comfort needs requires further analysis.

Figure 6 shows the analysis results of the range of thermal comfort thresholds in the absence of direct sunlight. The subjective percentage of dissatisfaction (PD), determined by the subjects' reports of their thermal comfort status, increased as the operative temperature decreased or increased. An optimal operative temperature range exists that meets thermal comfort needs. According to the authors of [27], if the 80% acceptability limits were used as the threshold for the indoor thermal comfortable environment, the allowable operative temperatures should not be lower than 21.0 °C or higher than 25.5 °C. In comparison with previous studies, this temperature range is consistent with the general rule of indoor thermal comfort in non-direct sunlight environments. This confirms that the changes in indoor thermal comfort are in line with the general laws of indoor thermal environments with the longwave radiation of windows alone.

3.2.2. Thermal Comfort under Direct Solar Radiation

The thermal effect of direct solar radiation on the human body is to enable thermal comfort to move in the direction of partial heat. As long as it is under the influence of solar radiation, the human body will passively absorb radiation heat. When exposed to direct sunlight at the same room temperature, the human body may be more prone to overheating. The experimental data were grouped based on the presence or absence of direct sunlight and statistically analyzed the subjective thermal sensation differences between the two groups of occupants at different operative temperatures at 0.5 °C intervals, as shown in Figure 7. The calculated operative temperature was the combined value of environmental radiant temperature and air temperature from longwave radiation heat alone.

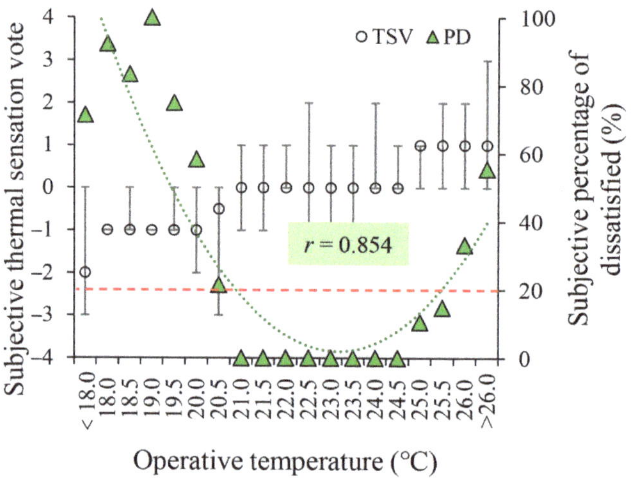

Figure 6. Threshold of allowable operative temperatures without solar radiation.

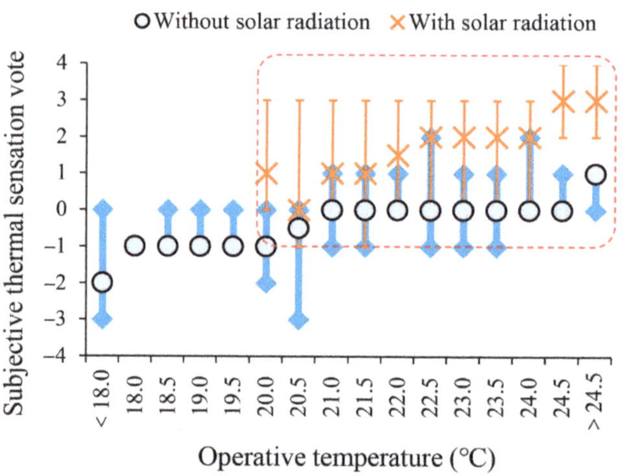

Figure 7. Comparison of subjective thermal sensations under direct and non-direct sunlight environments.

At the same operative temperature, a deviation in the thermal sensation existed between people exposed to direct sunlight and those without direct sunlight. The degree of deviation increased with an increase in the operative temperature. When exposed to direct sunlight, the human body is more likely to experience a feeling of overheating. As shown in the red-dashed area in Figure 7, when the operative temperature was greater than 24.5 °C, the subjective thermal sensation vote values of the two experimental groups reached an overall deviation of more than 2 units.

In Figure 8, the subjective thermal sensation vote (TSV) was compared with the predicted mean vote (*PMV*). It was indicated that the linear correlation between TSV and *PMV* was prominent when the occupants were not exposed to solar radiation, while the linear correlation was significantly weaker when solar radiation landed on the occupants. These orange data points were clustered in the upper left corner of the chart. It was suggested that the *PMV* would underestimate the thermal sensations of subjects exposed to solar radiation.

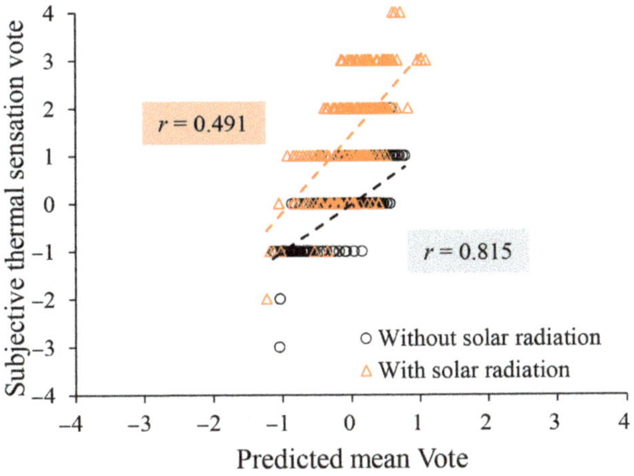

Figure 8. The relationship between subjective thermal sensation vote (TSV) and predicted mean vote (*PMV*) under direct and non-direct sunlight environments.

The analysis results of Figures 7 and 8 both indicate that it was no longer appropriate to use the conventional indoor thermal environment indices to assess thermal sensation when the human body was exposed to direct sunlight. The operative temperature only considered the effect of longwave radiation in conventional thermal comfort indices. In response to this issue, we adopted the SMRT method to modify the conventional operative temperature and considered the additional heating load on the human body under the influence of solar radiation as described in Section 2.6.3. The solar-adjusted operative temperature was linearly refitted with the subjective thermal sensation vote value, which was compared with the linear fitting results with conventional operative temperatures shown in Figure 9.

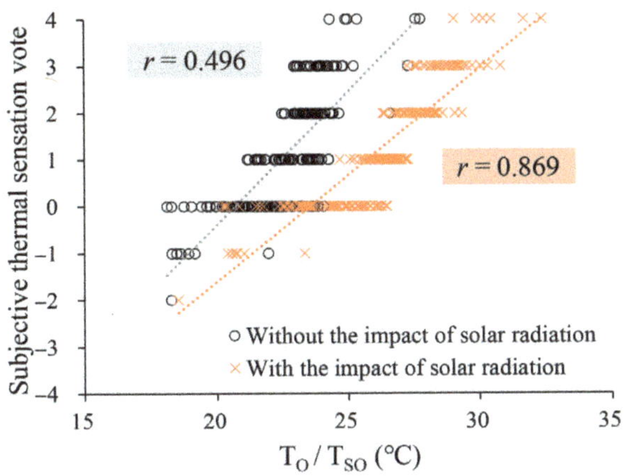

Figure 9. Comparison of subjective thermal sensation polling values and linear regression results of T_o and T_{so}.

The linear correlation between operative temperatures and subjective thermal sensations was 0.496, while the solar-adjusted linear correlation between the two was 0.876. This suggests that the solar-adjusted operative temperature can better predict the thermal sensation under the effect of solar radiation. The correlation between the operative temperatures and thermal sensations showed that the conventional calculation method of the mean radiant temperature underestimated the thermal sensation with solar radiation.

To further analyze the thermal comfort threshold of occupants under the effect of solar radiation, the solar-adjusted operative temperature was fitted to the thermal sensation as shown in Figure 10. If the 80% acceptability limits were used as the threshold for the indoor thermal comfortable environment, the solar-adjusted operative temperature should be controlled between 21.0 °C and 25.0 °C to maintain thermal comfort under the impact of solar radiation. Compared with Figure 6, the maximum operative temperature that occupants can withstand under the effect of solar radiation decreased. This suggests that when occupants are exposed to direct sunlight, lower indoor air temperatures are required to maintain a state of thermal comfort.

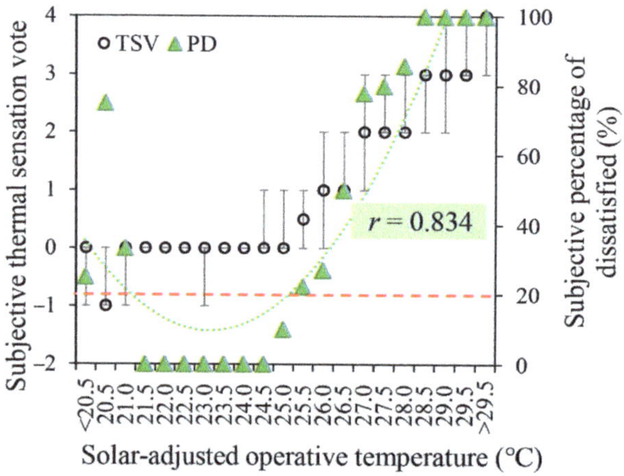

Figure 10. Threshold of allowable operative temperatures with solar radiation.

4. Conclusions

The impact of exterior windows on the indoor thermal environment and thermal comfort is not only reflected in air temperature but also in the uniformity of environmental radiant temperature and thermal environment. The study of the impact of exterior windows on the indoor thermal environment and thermal comfort can help to further improve the energy-saving design of windows in office buildings. In this study, field experiments were carried out to analyze the influence of windows on the indoor thermal environment in office buildings from the perspective of temperature fluctuations on the surface of the window glass and solar radiation transmission. The following conclusions were obtained:

- Outdoor air temperature and the intensity of outdoor solar radiation have a significant effect on the temperature change of the internal surface of windows. Changes in the surface temperature of the window trigger the fluctuation of the mean indoor radiant temperature and increase the non-uniformity of the indoor thermal environment. On a clear winter day, the fluctuation in the mean radiant temperature in the indoor near-window area can reach 7.8 °C, owing to the solar radiation;
- When the human body is not directly exposed to sunlight, the thermal comfort changes follow the general law under the action of longwave radiation from the window alone. When the human body is exposed to direct sunlight, the conventional indoor thermal environment evaluation indices underestimate the thermal sensation. The mean indoor radiant temperature must be corrected and the additional heat load caused by solar shortwave radiation should be considered;
- When the human body is directly exposed to sunlight, the threshold range of operative temperature to maintain thermal comfort is reduced. In this study, the maximum

operative temperature that the human body can withstand decreased by approximately 0.5 °C.

Windows play a critical role in building energy consumption and indoor thermal comfort. The outcomes of this study indicate that the design of exterior windows in office buildings should also take into account the effect of solar radiation transmission on thermal comfort, as well as the impact of temperature fluctuations on the window's inner surfaces. For the thermal performance of windows, reducing the heat-transfer coefficient of windows, as well as the absorption and re-conduction of solar radiation by glasses, is an effective way to weaken the impact of windows on the indoor thermal environment. Further, the shading devices, including the use of Venetian blinds and sun shields, which can reduce the transmission of shortwave radiation, can alleviate the occupant's discomfort. However, it should be noted that these measures can also lead to an increase in lighting energy consumption. These measures should balance the demand for thermal comfort and daylighting.

In terms of the design of air conditioning systems, the indoor air temperature of office buildings with large windows should be regulated in zones. For example, in the near-window area, it is necessary to reduce the supply air temperature when solar radiation reaches the occupants. Air temperature and radiant temperature both are important factors in the indoor thermal environment. Conventional practice is to control the air conditioning using temperature sensors that respond only to air temperatures. According to the results of this study, a sensor that responds to the combination of air and radiant temperatures may be more appropriate under the circumstances close to exterior windows.

In this study, potential differences between TSV experienced by males and females did not emerge, limited by the number of participants. This study focused primarily on the south façade and on the winter months when the solar radiation transmission through windows is most significant. The design of windows should balance the indoor thermal environment requirements throughout the year. In future research work, more orientation conditions should be considered and there is a need to cover different time periods, such as summer and transitional seasons, in order to provide a more comprehensive guide to the design of office building windows.

Author Contributions: Conceptualization, B.S. and L.B.; methodology, B.S.; validation, L.B. and L.Y.; investigation, B.S.; writing, B.S. and L.B.; supervision, L.Y. All authors have read and agreed to the published version of the manuscript.

Funding: This research was funded by the National Natural Science Foundation of China, grant number 52108093.

Data Availability Statement: Data is contained within the article.

Conflicts of Interest: The authors declare no conflicts of interest.

Nomenclature

CZ	Cold zone
HSCWZ	Hot summer and cold winter zone
PD	The subjective percentage of dissatisfaction
PMV	Predicted mean vote
SD	Standard deviation
SHGC	Solar heat gain coefficient
SMRT	Solar-adjusted mean radiant temperature
TSV	Subjective thermal sensation vote
U-value	The rate of transfer of heat through matter
F_{s-i}	The relative viewing angle between the human body and the ith enclosure
$F_{s\to j}$	The relative viewing angle between the human body and the jth window
I_b	The intensity of beam solar radiation, W/m^2

I_{dj}	The intensity of diffuse solar radiation of jth window, W/m^2
M	The number of windows
N	The number of enclosure surface
T_i	The surface temperature of the ith enclosure relative to the orientation of a human body, °C
T_o	The operative temperature
T_r	The mean radiant temperature
T_{r-long}	The mean radiant temperature of longwave radiation, °C
T_{smrt}	The solar-adjusted mean radiant temperature
T_{so}	The solar-adjusted operative temperature
X_i	The value of the ith sample in the X dataset
$\overline{X}$	The mean value of the X dataset
Y_i	The value of the ith sample in the Y dataset
$\overline{Y}$	The mean value of the Y dataset
f_p	Projected area factor
h_c	The heat − transfer coefficient by convection, $W/(m^2 \cdot °C)$
h_r	The heat − transfer coefficient by radiation, $W/(m^2 \cdot °C)$
r	The Pearson correlation coefficient
t_a	Air temperature
α_{irr}	Relative absorption coefficient referring to the solar radiation
ε_S	Emissivity of the subject
σ	Stefan–Boltzmann constant

References

1. Building Energy Efficiency Research Center. *Annual Development Research Report of China Building Energy Efficiency*; China Architecture & Building Press: Beijing, China, 2022; pp. 14–30.
2. Lim, Y.W.; Kandar, M.Z.; Ahmad, M.H.; Ossen, D.R.; Abdullah, A.M. Building façade design for daylighting quality in typical government office building. *Build. Environ.* **2012**, *57*, 194–204. [CrossRef]
3. Troup, L.; Phillips, R.; Eckelman, M.J.; Fannon, D. Effect of window-to-wall ratio on measured energy consumption in US office buildings. *Energy Build.* **2012**, *203*, 109434. [CrossRef]
4. Cuce, E.; Riffat, S.B. A state-of-the-art review on innovative glazing technologies. *Renew. Sustain. Energy Rev.* **2015**, *41*, 695–714. [CrossRef]
5. Amaral, A.R.; Rodrigues, E.; Gaspar, A.R.; Gomes, Á. A thermal performance parametric study of window type, orientation, size and shadowing effect. *Sustain. Cities Soc.* **2016**, *26*, 456–465. [CrossRef]
6. Kaasalainen, T.; Mäkinen, A.; Lehtinen, T.; Moisio, M.; Vinha, J. Architectural window design and energy efficiency: Impacts on heating, cooling and lighting needs in Finnish climates. *J. Build. Eng.* **2020**, *27*, 100996. [CrossRef]
7. Kim, S.S.; Bae, M.J.; Kim, Y.D. Policies and status of window design for energy efficient buildings. *Procedia Eng.* **2016**, *146*, 155–157. [CrossRef]
8. Lang, S. Progress in energy-efficiency standards for residential buildings in China. *Energy Build.* **2004**, *36*, 1191–1196. [CrossRef]
9. Höppe, P.R. Indoor climate. *Experientia* **1993**, *49*, 775–779. [CrossRef]
10. Wang, T.P.; Wang, L.B.; Li, B.Q. A model of the long-wave radiation heat transfer through a glazing. *Energy Build.* **2013**, *59*, 50–61. [CrossRef]
11. Halawa, E.; van Hoof, J.; Soebarto, V. The impacts of the thermal radiation field on thermal comfort, energy consumption and control—A critical overview. *Renew. Sustain. Energy Rev.* **2014**, *37*, 907–918. [CrossRef]
12. Yang, R.; Zhang, H.; You, S.; Zheng, W.; Zheng, X.; Ye, T. Study on the thermal comfort index of solar radiation conditions in winter. *Build. Environ.* **2020**, *167*, 106456. [CrossRef]
13. Ge, H.; Fazio, P. Experimental investigation of cold draft induced by two different types of glazing panels in metal curtain walls. *Build. Environ.* **2004**, *39*, 115–125. [CrossRef]
14. Gan, G. Analysis of mean radiant temperature and thermal comfort. *Build. Serv. Eng. Res. Technol.* **2001**, *22*, 95–101. [CrossRef]
15. Chaiyapinunt, S.; Phueakphongsuriya, B.; Mongkornsaksit, K.; Khomporn, N. Performance rating of glass windows and glass windows with films in aspect of thermal comfort and heat transmission. *Energy Build.* **2005**, *37*, 725–738. [CrossRef]
16. Sengupta, J.; Chapman, K.S.; Keshavarz, A. Window performance for human thermal comfort. *ASHRAE Trans.* **2005**, *111*, 254–275.
17. Sengupta, J.; Chapman, K.S.; Keshavarz, A. Development of a methodology to quantify the impact of fenestration systems on human thermal comfort. *ASHRAE Trans.* **2005**, *111*, 239–253.
18. Zhao, K.; Gui, X.C.; Ge, J. Influence of solar radiation on thermal comfort in lager spaces and corresponding design of indoor paeameters. *Acta Energiae Solaris Sin.* **2019**, *40*, 2655–2662.

19. Chaiyapinunt, S.; Khamporn, N. Effect of solar radiation on human thermal comfort in a tropical climate. *Indoor Built Environ.* **2021**, *30*, 391–410. [CrossRef]
20. Arens, E.; Gonzalez, R.; Berglund, L. Thermal comfort under an extended range of environmental conditions. *ASHRAE Trans.* **1986**, *92*, 18–26.
21. Schutrum, L.F.; Stewart, J.L.; Nevins, R.G. A subjective evaluation of effects of solar radiation and reradiation from windows on the thermal comfort of women. *ASHRAE Trans.* **1968**, *74*, 115–128.
22. Lyons, P.; Arates, D.; Huizenga, C. Window performance for human thermal comfort. *ASHRAE Trans.* **1999**, *73*, 400–420.
23. Bessoudo, M.; Tzempelikos, A.; Athienitis, A.K.; Zmeureanu, R. Indoor thermal environmental conditions near glazed facades with shading devices–Part I: Experiments and building thermal model. *Build. Environ.* **2010**, *45*, 2506–2516. [CrossRef]
24. Tzempelikos, A.; Bessoudo, M.; Athienitis, A.K.; Zmeureanu, R. Indoor thermal environmental conditions near glazed facades with shading devices–Part II: Thermal comfort simulation and impact of glazing and shading properties. *Build. Environ.* **2010**, *45*, 2517–2525. [CrossRef]
25. *GB 50189-2015*; Design Standard for Energy Efficiency of Public Buildings. 1st ed. China Architecture & Building Press: Beijing, China, 2015; pp. 17–18.
26. *GB 50736-2012*; Design Code for Heating Ventilation and Air Conditioning of Civil Buildings. 1st ed. China Architecture & Building Press: Beijing, China, 2012; pp. 4–10.
27. *GB/T 50785-2012*; Evaluation Standard for Indoor Thermal Environment in Civil Buildings. 1st ed. China Architecture & Building Press: Beijing, China, 2012; pp. 3–15.
28. Bai, L.; Song, B.; Yang, L. Developing the New Thermal Climate Zones of China for Building Energy Efficiency Using the Cluster Approach. *Atmosphere* **2022**, *13*, 1498. [CrossRef]
29. *ANSI/ASHRAE Standard 55*; Thermal Environmental Conditions for Human Occupancy. 1st ed. American Society of Heating, Refrigerating and Air-Conditioning Engineers (ASHRAE): Atlanta, GA, USA, 2017; pp. 7–20.
30. Pantavou, K.; Koletsis, I.; Lykoudis, S.; Melas, E.; Nikolopoulou, M.; Tsiros, I.X. Native influences on the construction of thermal sensation scales. *Int. J. Biometeorol.* **2020**, *64*, 1497–1508. [CrossRef]
31. ISO. *Ergonomics of the Thermal Environment Instruments for Measuring Physical Quantities*, 1st ed.; International Organization for Standards: Geneva, Switzerland, 1998; pp. 31–34.
32. Singh, M.C.; Garg, S.N.; Jha, R. Different glazing systems and their impact on human thermal comfort—Indian scenario. *Build. Environ.* **2008**, *43*, 1596–1602. [CrossRef]
33. La Gennusa, M.; Nucara, A.; Rizzo, G.; Scaccianoce, G. The calculation of the mean radiant temperature of a subject exposed to the solar radiation—A generalised algorithm. *Build. Environ.* **2005**, *40*, 367–375. [CrossRef]
34. La Gennusa, M.; Nucara, A.; Pietrafesa, M.; Rizzo, G. A model for managing and evaluating solar radiation for indoor thermal comfort. *Sol. Energy* **2007**, *81*, 594–606. [CrossRef]
35. Marino, C.; Nucara, A.; Pietrafesa, M.; Polimeni, E. The effect of the short wave radiation and its reflected components on the mean radiant temperature: Modelling and preliminary experimental results. *J. Build. Eng.* **2017**, *9*, 42–51. [CrossRef]
36. Khamporn, N.; Chaiyapinunt, S. Effect of installing a venetian blind to a glass window on human thermal comfort. *Build. Environ.* **2014**, *82*, 713–725. [CrossRef]
37. Zhang, H.; Yang, R.; You, S.; Zheng, W.; Zheng, X.; Ye, T. The CPMV index for evaluating indoor thermal comfort in buildings with solar radiation. *Build. Environ.* **2018**, *134*, 1–9. [CrossRef]
38. Huang, L.; Zhai, Z.J. Critical review and quantitative evaluation of indoor thermal comfort indices and models incorporating solar radiation effects. *Energy Build.* **2020**, *224*, 110204. [CrossRef]
39. Cannavale, A.; Ayr, U.; Fiorito, F.; Martellotta, F. Smart electrochromic windows to enhance building energy efficiency and visual comfort. *Energies* **2020**, *13*, 1449. [CrossRef]
40. Huang, Y.; Niu, J.L. Energy and visual performance of the silica aerogel glazing system in commercial buildings of Hong Kong. *Constr. Build. Mater.* **2015**, *94*, 57–72. [CrossRef]
41. Wang, S.; Zhou, Y.; Jiang, T.; Yang, R.; Tan, G.; Long, Y. Thermochromic smart windows with highly regulated radiative cooling and solar transmission. *Nano Energy* **2021**, *89*, 106440. [CrossRef]
42. Zhang, Y.; Zhao, R. Relationship between thermal sensation and comfort in non-uniform and dynamic environments. *Build. Environ.* **2009**, *44*, 1386–1391. [CrossRef]

energies

MDPI

Article

Research on Multi-Objective Optimization Design of University Student Center in China Based on Low Energy Consumption and Thermal Comfort

Ming Liu [1], Yufei Que [1], Nanxin Yang [1], Chongyi Yan [2,*] and Qibo Liu [1,3,*]

[1] School of Architecture, Chang'an University, Xi'an 710061, China; liuming@chd.edu.cn (M.L.); 2020903783@chd.edu.cn (Y.Q.); 2022241002@chd.edu.cn (N.Y.)
[2] CCCC First Highway Consultant Co., Ltd., Xi'an 710068, China
[3] Engineering Research Center of Highway Infrastructure Digitalization, Ministry of Education, Xi'an 710064, China
* Correspondence: 15229324203@163.com (C.Y.); lucy@chd.edu.cn (Q.L.); Tel.: +86-15229324203 (C.Y.); +86-13572508860 (Q.L.)

Abstract: Ensuring optimal building performance is vital for enhancing student activity comfort and fostering energy-saving initiatives toward low-carbon objectives. This paper focuses on university student centers in China, aiming to diminish building energy consumption while enhancing indoor thermal comfort. Parametric modeling of typical cases is executed using the Grasshopper 1.0.0007 software package, and the simulation of building energy consumption and indoor thermal comfort relies on the Ladybug and Honeybee plug-in. Employing a multi-objective optimization design method and the Octopus multi-objective optimization algorithm, this study integrates numerical simulations and on-site surveys to analyze how factors like building form, orientation, envelope structure, and others impact the indoor and outdoor environment. A comprehensive optimization design approach is implemented for the building's exterior components, including the walls, windows, roof, and shading system. After conducting a comparative analysis of the annual comprehensive energy consumption and indoor thermal comfort before and after the optimization plan, it is determined that implementing these measures reduces the annual comprehensive energy consumption of the building under study by 58.8% and extends the duration of indoor thermal comfort by 53.0%. This study presents a practical optimization design methodology for university student center architecture in China, aiding architects in decision making and advocating for energy-efficient building designs.

Keywords: student center; low energy consumption; thermal comfort; multi-objective optimization design; numerical simulation

1. Introduction

The global construction industry is shifting from making quantitative changes to focusing on qualitative transformations, particularly in university construction, where there is an emphasis on fostering the sustainable development of green campuses. By prioritizing design goals centered around durability, livability, and low carbon emissions, solutions have been identified for addressing challenges encountered by different types of university buildings associated with high energy consumption [1]. Within this context, the student center, known for its diverse and intricate functional spaces, becomes a critical focal point in striving to achieve low energy consumption across the campus.

In constructing a student center, a comprehensive approach to considering thermal comfort and green energy efficiency within indoor environments is crucial. On the one hand, maintaining an appropriate temperature and humidity directly impacts students' utilization and frequency of space. On the other hand, some student centers sacrifice high energy consumption to sustain indoor thermal comfort, resulting in increased energy usage. Hence, designing an energy-saving optimization scheme for the student center demands a multi-objective optimization study focusing on achieving both low energy consumption and optimal indoor thermal comfort. This approach aims to drive the development of energy-efficient campuses across China.

2. Literature Review

In recent times, scholars have conducted multi-objective optimization research concerning low building energy consumption and thermal comfort. Their primary focus has been on examining the interaction between optimization algorithms and simulation tools.

Regarding optimization algorithms, several studies have been conducted. Caldas et al. focused on office buildings, considering factors associated with different orientations and exterior windows as variables, with total building energy consumption as the optimization objective [2]. Magnier examined building equipment and envelope structure as variables, utilizing the Non-dominated Sorting Genetic Algorithm-II (NSGA-II) algorithm combined with the ANN algorithm to target building energy consumption and thermal comfort as optimization objectives [3]. Yuan Fang et al. centered on the design of external windows in office buildings. They simulated building performance using the Grasshopper platform and Ladybug and Honeybee plug-in, analyzing the impact of these windows on energy consumption and proposing relevant optimization strategies [4]. Li Haiquan proposed a multi-objective building energy-saving model based on a genetic algorithm, focusing on parameters such as body shape coefficient and windward area [5]. Benedek Kiss explored a multi-objective optimization technique, specifically the Direct MultiSearch method, applied to a case study involving a multi-story residential building [6]. Giouri ED investigated the exterior windows of high-rise office buildings, analyzing their influence on energy consumption and proposing corresponding optimization strategies [7]. Badr Chegari utilized a multi-objective optimization method based on an intelligent agent model, employing machine learning tools like ANN and the MOPSO algorithm for optimization [8]. Facundo Bre introduced an effective method involving metamodels for solving multi-objective optimization problems in building performance, combining NSGA-II with an artificial neural network (ANN) metamodel [9]. Aiman Albatayneh conducted a sensitivity analysis (SA) using the Design Builder software Version 6.1 package, assessing 12 design variables simultaneously for their impact on heating and cooling loads through regression methods [10].

Regarding the interaction of simulation tools, Ehsan Asadi investigated existing external wall insulation and other envelope structures as components. They considered building energy consumption and thermal comfort as optimization objectives and conducted research using GenOpt and MATLAB Version 8.0 [11]. Erlendsson focused on the effective lighting level within residential atrium spaces. They utilized the Grasshopper platform and Honeybee plug-in to study the impact of glass size and material in atriums [12]. E. Belloni and colleagues investigated a novel aerogel glazing system (AGS) employing a simulation model to calculate the annual energy demand of a reference building in Tokyo. Through building performance simulation, it was demonstrated that AGS could reduce energy requirements in cold climates, concurrently decreasing the consumption associated with heating, cooling, and lighting [13]. Benedek Kiss introduced a modular parametric optimization framework for a multi-apartment building. The framework has key components including geometry, shell structure, installation, and heating energy. It integrates advanced building modeling, life cycle assessment (LCA), energy calculations, and environmental optimization methods, both single and multi-objective [14]. Zhang Anxiao's research centered on primary and secondary schools in cold areas. They aimed to optimize building energy consumption and room thermal comfort, employing multi-objective optimization theory and algorithms to verify the feasibility of a multi-objective optimization platform [15]. Yang Zhao and associates conducted a study on the thermal comfort of semi-outdoor spaces within a university building in Guangzhou, China, utilizing questionnaires for evaluation. The results revealed that both air temperature and mean radiant temperature within the courtyards were notably elevated. Planting trees in these courtyards could reduce the Physiological Equivalent Temperature (PET) from "very hot" to "hot", thereby aiding in the improvement of thermal comfort [16]. Erminia Attaianese and her team proposed an integrated method for creating sustainable indoor architectural environments, emphasizing Indoor Environmental Quality (IEQ) and overall building quality. They validated this approach through a subjective survey carried out at the University of Salerno's Fisciano Campus in Italy [17]. Liu Qianqian's study considered envelope structure factors as optimization variables, focusing on building energy consumption, lighting environment, and renovation economic cost. They utilized Rhino and Grasshopper for multi-objective optimization, examining objective optimization during the design phase of neural networks [18]. Table 1 presents a more systematic representation of the literature on achieving low building energy consumption and enhancing thermal comfort.

Table 1. Representative literature on building low energy consumption and thermal comfort.

Optimization Objectives	Research Object	Methodology	Literature
Building energy consumption and thermal comfort	Building equipment and envelope	Optimization algorithm	[2,11,15,19,20]
	Primary and secondary schools in cold climates	Interaction of simulation tools + optimization algorithm	[13,21]
Building energy consumption	Orientation and exterior window form of office building	Interaction of simulation tools	[1,3,6,16,22]
	Sensitivity analysis of 12 design variables	Interaction of simulation tools	[10]
	Factors for envelope	Interaction of simulation tools + optimization algorithm	[14,17,23]
Building energy efficiency	Apartment	Interaction of simulation tools	[7,24]
	Shape factor and windward area	Interaction of simulation tools + optimization algorithm	[4]
	Atrium space	Interaction of simulation tools	[11,12]
Multi-objective optimization tool	Multi-story residential building	Optimization algorithm	[5]
	Multi-objective optimization of building performance		[9,21,25–28]
	Building optimization based on intelligent algorithm		[8,29–31]
	Building optimization with machine learning		[32–34]

Upon reviewing existing research, several key observations emerge:

(1) In the realm of architectural optimization design, intelligent parametric design and optimization algorithms have progressively become commonly utilized methods in architecture. This is particularly evident in the comprehensive platforms based on Grasshopper.

(2) Using multi-objective optimization to improve building energy efficiency mainly centers on integrating various algorithms to optimize building structure, design, envelope, and equipment. This approach aims to enhance both energy efficiency and overall comfort levels.

(3) Research on integrating parametric building design and performance simulation is limited. More specifically, there is a shortage of studies that utilize multi-objective optimization algorithms within the context of campus public buildings, particularly when examining student centers.

Drawing from the aforementioned existing research, this study focuses on the student center in China. The research leverages the Grasshopper platform for building performance simulation analysis alongside a multi-objective optimization module to analyze the current state and outcomes of computer simulations. The aim is to identify existing issues in the usage of university student centers, considering key evaluation criteria for energy-saving design in public buildings. Subsequently, the study focuses on two primary goals: reducing building energy consumption and enhancing indoor thermal comfort. It reforms and optimizes the main factors influencing energy conservation in buildings. The innovation of this study lies in integrating the establishment of parameterized building models, building performance analysis, and multi-objective algorithm optimization on the same platform, avoiding errors in data exchange and repeated modeling between traditional platforms. Comparing the results of the optimization scheme with the initial simulation data of the building model, a comprehensive analysis is conducted to derive energy-saving optimization strategies for the student center. This study introduces fresh insights and theoretical references for architectural design methodologies specific to university student centers, considering regional climate characteristics in China.

3. Methodology

3.1. Morphological Characteristics of University Student Centers in China

Alongside the continuous evolution of modern college education concepts, the student center has become an integral part of campuses, drawing widespread attention due to its distinctive architectural characteristics and functionalities. Referring to pertinent content in "China's Architectural Design Data Set" (3rd edition) [35], functional types are summarized in Table 2. Given the inclusive and diverse nature of student centers, organizing their interior spaces must not only ensure the autonomy of each functional room and the distinction between active and quiet areas but also prioritize flexibility in usage and the integration of various functions. Following the guidelines outlined in "Code for the Architectural Design of Cultural Centers" JGJ/T 41-2015 [36] and relevant specifications in the Architectural Design Data Set (3rd edition), plans are primarily categorized into three types (centralized combined type, series combined type, and courtyard enclosed type), as illustrated in Table 3.

Table 2. Program of student center recommended by the Architectural Design Data Set [35].

Name		Program
Office		Student union, science and technology service center, information service center, psychological counseling, career center, etc.
Event and management occupancy	General club room	Literature club, photography club, fine arts club, etc.
	Literary club rehearsal room	Vocal, dance, etc.
	Group activity	Rooms for exhibitions, discussions, lectures, fellowship, and other activities organized by student organizations or individuals
	Recreation	Board, billiards, video games, etc.
	Fitness	gym
Hall		Professional stage, can be large-scale performance places
Multi-purpose room		Ballroom, a venue for larger events such as exhibitions, assemblies, or rehearsals
Others	Broadcasting station	Broadcasting station
	Service room	Bookstores, banks, haircuts, photocopying, repairs, etc.
	Small food and beverage occupancy	Fast food restaurant, cafe, etc.

Although factors such as geographical location, environment, university scale, and construction period vary among colleges and universities, the functional types of student centers across these institutions are fundamentally similar. However, spatial organization is influenced by geographical conditions and climate characteristics. This study employs a literature review, map consultation, and site visits. Several student centers in Beijing, Tianjin, Nanjing, Guangzhou, Xi'an, and other cities were investigated. Due to space constraints, Table 4 presents eight representative university student centers in China.

Table 3. Organizational form of plan [35,36].

Classification	(a) Centralized Combined Type	(b) Series Combined Type	(c) Courtyard Enclosed Type
Legend  roo fove hall	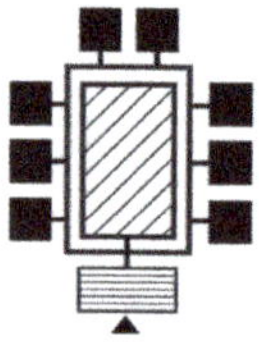	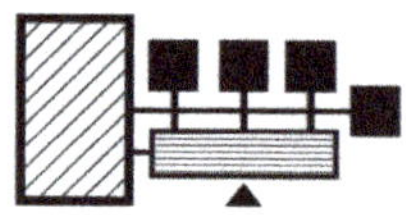	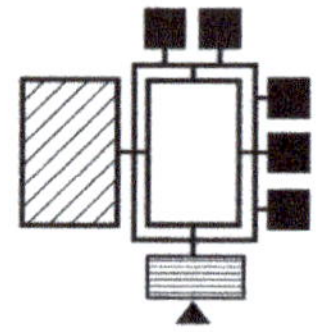

Table 4. Program analysis diagram of student center in China.

Region (South)	Layout and Functional Zoning	Region (North)	Layout and Functional Zoning
Student center, South China University of Technology	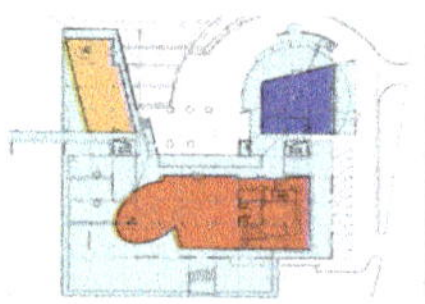	Student center, Tsinghua University	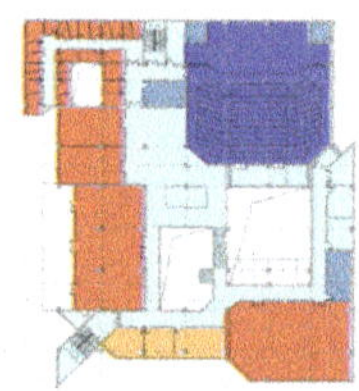
Student center, Shenzhen University	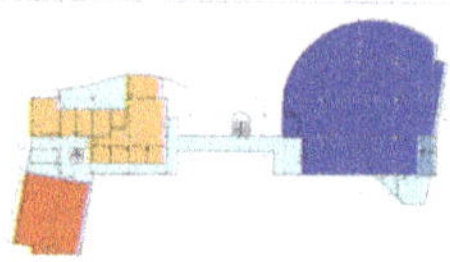	Student center, Tianjin Polytechnic University	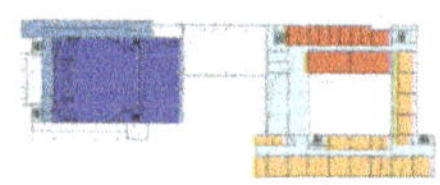
Student center, Southeast University	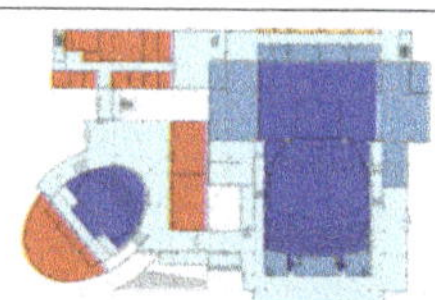	Student center, Lanzhou University	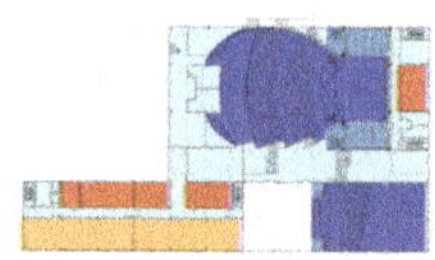
Student center, Nanjing University	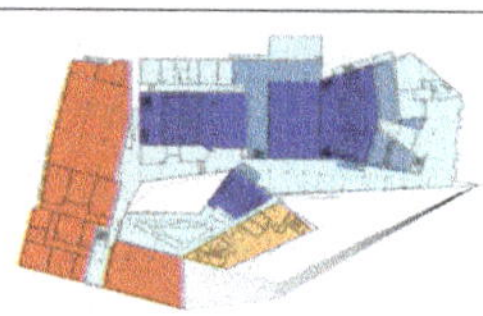	Student center, Chang'an University	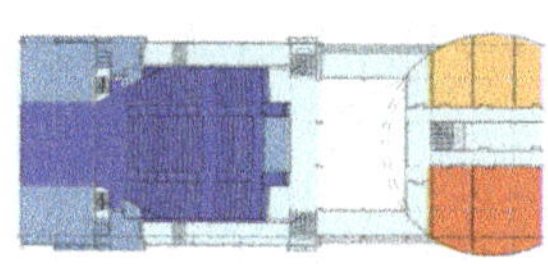
office; event and management occupancy; hall; multi-purpose room; others			

Through investigation, we observed distinctive characteristics in the function and utilization of the student center, primarily manifested as follows:

(1) Functional diversity: The student center integrates educational, social, and recreational functions, offering adaptable space to address diverse needs.

(2) Layout response to climate: As depicted in Table 4, student centers in southern regions predominantly adopt series and courtyard layouts, while in the north, centralized layouts prevail.

(3) Integration of spaces in varied scales: The design of the student center encompasses extensive areas such as halls, medium-sized spaces like activity and rehearsal rooms, and smaller spaces including offices and service rooms. The combination, utilization, and management of these spaces are notably intricate.

(4) Unique utilization patterns: The peak usage of student centers differs significantly from teaching buildings; it tends to concentrate more during after-school hours, including nights and non-class periods, and particularly on weekends.

3.2. *Typical Case Study*

This study chose a university student center located in Xi'an, Shaanxi Province, northern China for several reasons:

(1) The Xi'an region is situated in a cold zone with a warm temperate continental monsoon climate, characterized by distinct cold, warm, dry, and wet seasons. Buildings in this area must meet insulation design requirements and consider natural ventilation and shading design. Additionally, Xi'an is home to numerous universities with extensive building areas. The prevalence of high energy consumption without adequate indoor thermal comfort maintenance, particularly among major universities in the area, underscores the need for research and optimization of the thermal environment in academic buildings. Findings from this region can provide valuable insights for buildings in other cold regions.

(2) China has a substantial inventory of "non-green" buildings characterized by low energy utilization and high operating energy consumption. With the introduction of the "dual-carbon" target, the green and energy-saving renovation of existing buildings is poised to become the prevailing trend in China's construction industry.

(3) Research reveals that some previously constructed student center buildings were constrained by factors such as outdated building energy efficiency standards, materials, technology, and equipment. Issues such as obsolete heating equipment, aging power lines and envelope structures, severe condensation, mold growth in indoor corners, and peeling of inner wall finishes have become increasingly prominent. The inadequate thermal performance of envelope structures results in indoor heat loss, elevated building energy consumption, and challenges in maintaining comfortable indoor temperatures.

(4) The existing research literature predominantly focuses on the spatial design of student centers, neglecting building performance and physical environment optimization design. Therefore, this study aims to explore energy-saving optimization design concepts and plans for this building type. The findings will facilitate improved operation, construction, and maintenance of student center buildings.

This study conducted a large-scale survey on student centers in universities in Xi'an, and selected 5 representative student centers constructed after 2000 as research cases for detailed investigation. The reason for selecting cases after 2000 is that during the investigation, it was found that some older student centers have been abandoned or repurposed, and lack good renovation value. These 5 universities represent the disciplines of liberal arts, science, and engineering in the Xi'an region, include Chang'an University (CHD), Xi'an International Studies University (XISU), Xi'an University of Finance and Economics (XUFE), Shaanxi Normal University (SNNU), and Xi'an University of Architecture and Technology (XAUAT). An overview of the case study buildings is provided in Table 5.

Table 5. Basic information of the five student centers.

Typical Cases	Year	Plan	Story	Floor Height	Gross Floor Area
CHD student center Before Design Standard for Energy Efficiency of Public Buildings GB50189-2005	2004	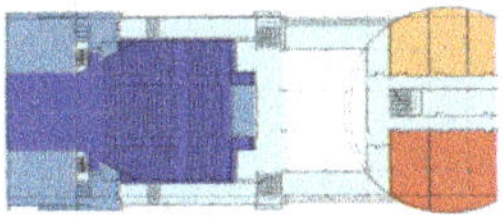	3	4.8 m	10,700 m^2

Table 5. *Cont.*

Typical Cases	Year	Plan	Story	Floor Height	Gross Floor Area
XISU student center Before Design Standard for Energy Efficiency of Public Buildings GB50189-2015	2009	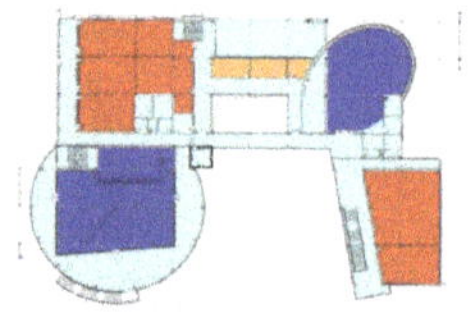	2	4.6 m	9156 m^2
SNNU student center Before Design Standard for Energy Efficiency of Public Buildings GB50189-2015	2011		4	4.2 m	9989 m^2
XUFE student center	2016	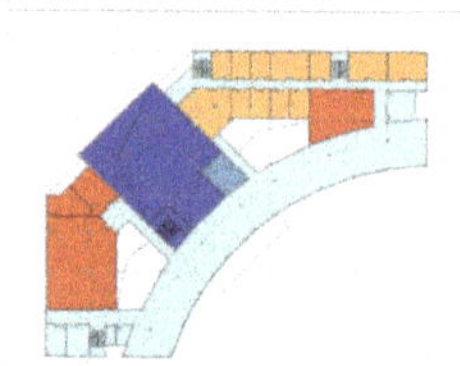	2	3.9 m	5231 m^2
XAUAT student center	2018	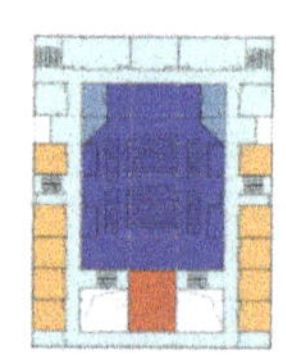	4	4.5 m	10,500 m^2

office; event and management occupancy; hall; multi-purpose room; others

We visited the management departments of five universities and obtained the completion drawings of the student centers, along with the total electricity consumption data of the buildings from 2017 to 2020. We discovered that all the student centers operate upon central air conditioning for cooling (electricity consumption) indoors in summer and centralized heating (natural gas) for indoor spaces in winter. It is observed that the total power consumption and power consumption per unit area of the student centers have consistently increased annually over the past four years, as depicted in Figures 1 and 2. These buildings, due to their comprehensive nature and combination of variously sized spaces, demonstrate low continuity in space utilization. The interior space of the building presents distinct usage requirements based on different functional divisions. During the construction phase, energy conservation considerations were not given significant priority. Combined with the effects of aging and wear during the building's use, it currently experiences poor thermal comfort and high energy consumption. Consequently, it fails to meet present-day requirements for energy conservation and consumption reduction, posing challenges in maintaining a comfortable indoor environment.

Considering the characteristics of the student center outlined above, including layout, spatial design, and usage features, and in conjunction with the fact that China issued the "Design Standard for Energy Efficiency of Public Buildings" GB50189-2005 [37] in 2005, this study ensures the comprehensiveness and integrity of relevant data. As a result, the CHD student center is chosen as the specific research subject. Despite being constructed in 2004, the building lacked comprehensive energy-saving measures during construction.

It presents a non-energy-efficient structure with a simplistic envelope design, inadequate thermal insulation, and numerous energy-related issues.

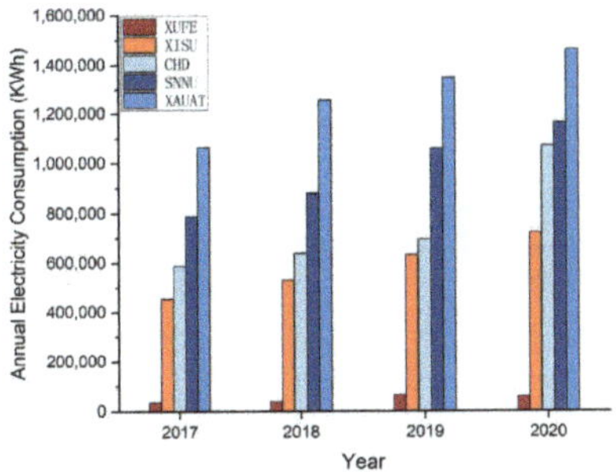

Figure 1. Total energy consumption of 5 student center buildings from 2017 to 2020.

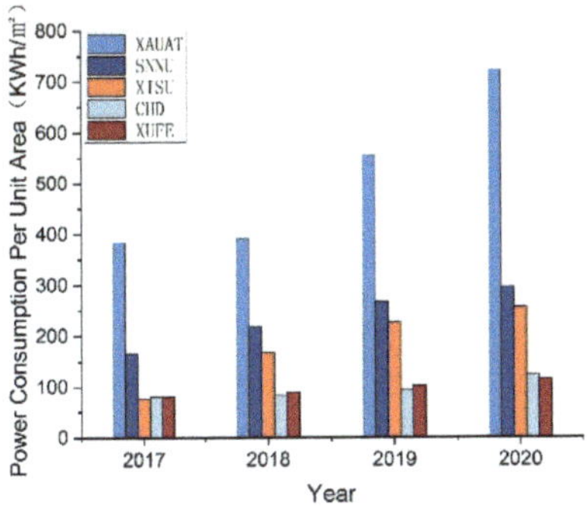

Figure 2. Electricity consumption per unit area of 5 student centers from 2017 to 2020.

3.3. Numerical Simulation

3.3.1. Setting Parameters for Typical Case Model

To examine the correlation between the building design of the student center and its impact on energy consumption and thermal comfort, this study utilized the Design Builder software package to simulate the case study. Climate data and building parameters were employed to parameterize elements such as building form, orientation, envelope structure, and other factors [9]. Simulation parameters were established following the calculation guidelines for physical properties of commonly used building materials, as outlined in Appendix B of the "Code for Thermal Design of Civil Buildings" GB 50176-2016 [38], and were based on the construction drawing practices of existing typical case models. Notably, the CHD Student Center lacks insulation material in its exterior walls and roof.

3.3.2. Basic Parameters of Heat Source in the Room

The model parameters were set in line with the actual characteristics of the case building, encompassing metrics such as the per capita area and room occupancy rate, the energy consumption rate of lighting and equipment per unit, lighting power density, switching times, and operational periods of the building's air conditioning and heating system [39]. Specific configurations are outlined as follows:

(1) Per capita area and occupancy: The area per person is set at 6 m^2; activity energy is 70 W; the thermal resistance unit of clothing is specified as 'clo', the occupants can adjust clothing according to their thermal feelings. Based on the Evaluation Standard for Indoor Thermal Environment in Civil Buildings GB/T 50785-2012, according to normal office clothing, when occupants wear thin pans and short-sleeve shirts, clo = 0.50 in summer;

when occupants wear long-sleeve shirts, trousers, warm jackets, and long sleeved and pants underwear, clo = 1.2 in winter [39,40].

(2) Lighting and equipment energy consumption per unit area indoors: Lighting power density is 9 W/m^2, triggering lamps when the working surface illumination falls below 500 lx. Fresh air volume is 30 $[m^3/(h)]$, and electrical equipment power density is 15 W/m^2.

(3) Operational schedule for public university buildings: Electrical equipment, room occupancy, and lighting have flexible timings, operating 7 days a week from 8:00 am to 10:00 pm based on the University's activity center usage. Lamp operation is limited to this period.

(4) Air conditioning and heating system operation: The cooling temperature is set to 26 °C, and the heating temperature to 18 °C [41]. Fan operation is synchronized with the student center's schedule, and inactive during other times. The study focused solely on indoor energy consumption and thermal comfort, excluding fan and coil models that might affect energy usage. Annual energy consumption calculations encompass cooling, heating, lighting, and other equipment.

After conducting field investigations on student centers in Xi'an, a representative typical building case was selected for parametric modeling and subsequent building performance simulation. Considering the climate characteristics of Xi'an, simulation analyses were conducted to assess building energy consumption and indoor thermal comfort. These simulation outcomes serve as a comparative data set for the optimization design scheme detailed in the subsequent paper. Illustrated in Figure 3 is a schematic diagram depicting the process and platform for building performance analysis.

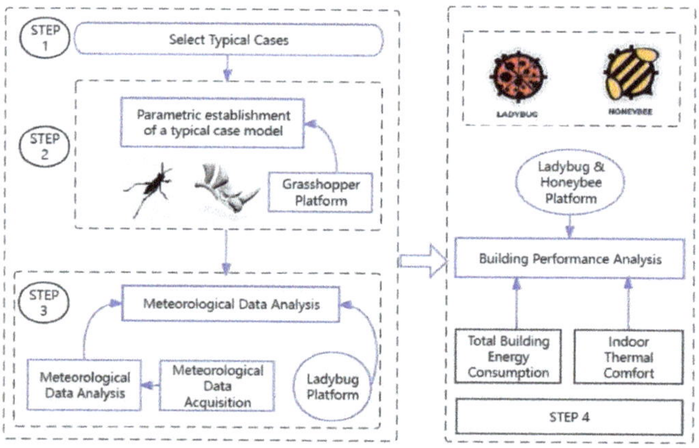

Figure 3. Schematic diagram of building performance analysis process and technology platform of student center.

3.4. *Simulation of Energy Consumption and Thermal Comfort*

3.4.1. Accuracy Verification of Typical Cases

Before conducting a building performance simulation, it is crucial to validate the accuracy of the model and the software's parameter settings. In this study, the measured outdoor temperature and humidity in Xi'an on 6 July 2021, were compared with the software-simulated data for verification. This comparison is depicted in Figures 4–6.

By organizing the measured data from outdoor measuring points, the results of outdoor simulations were compared against them. To simplify calculations, some simulation parameters were set to an ideal state. Consequently, the simulated outdoor humidity measurement points may register higher than the actual measurements, but they consistently reflect the trends of each measurement point. Upon comparison, the disparity between simulated temperature and humidity values and the measured values at various times of the day is less than 10%. This indicates that the software simulation results are largely in line with the test results, validating the feasibility of the software model and parameter settings.

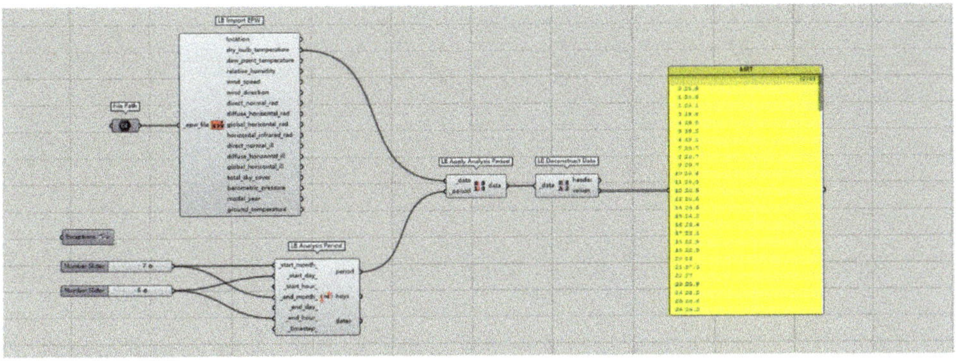

Figure 4. Establishment of outdoor hourly temperature platform on July 6.

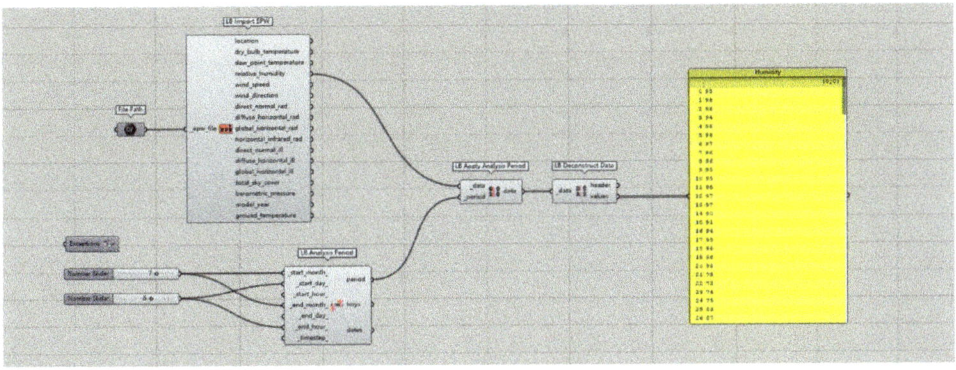

Figure 5. Establishment of outdoor hourly humidity platform on July 6.

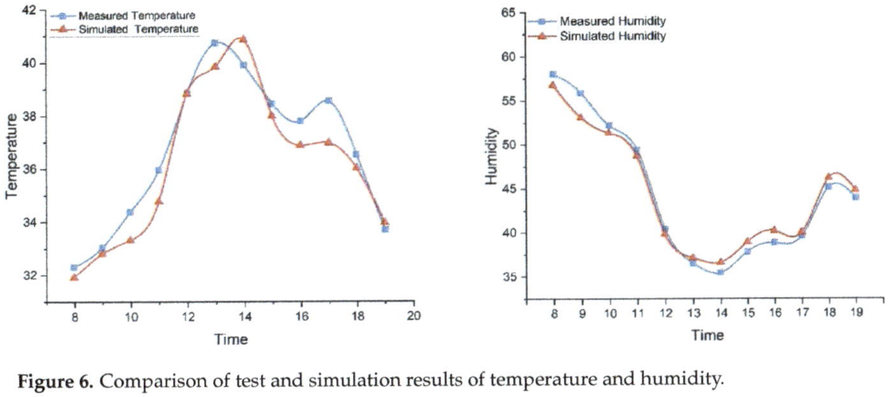

Figure 6. Comparison of test and simulation results of temperature and humidity.

3.4.2. Simulation Analysis of Energy Consumption

In this study, the Ladybug and Honeybee plug-in is used to calculate annual heating energy consumption, cooling energy consumption, artificial lighting energy consumption, and other equipment energy consumption to obtain annual comprehensive building energy consumption and comprehensive energy consumption per unit area. The visual results of building energy consumption and its data are depicted in Figure 7.

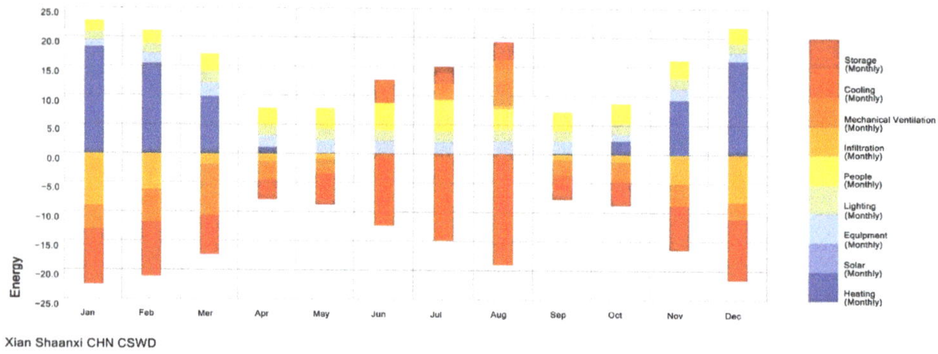

Figure 7. Energy consumption balance diagram of CHD Student Center.

As depicted in Figure 7, the annual energy consumption analysis of the CHD Student Center reveals a total energy consumption of 183.57 $kW{\cdot}h/m^2$ per unit area, with an accumulated annual heat consumption of 82.15 $kW{\cdot}h/m^2$. The primary cause for this lies in the aging of the envelope structure and excessive heat transfer coefficient. During winter, the primary heat loss occurs through the envelope, while in summer, air conditioning is predominantly used for interior comfort. Addressing this issue requires further optimization of the envelope and shade design, particularly for summer conditions.

3.4.3. Simulation Analysis of Indoor Thermal Comfort in Typical Buildings

The PMV-PPD index chosen for this study reflects the indoor thermal comfort within buildings. The visualized results and data illustrating PMV for indoor thermal comfort throughout the year are presented in Figure 8.

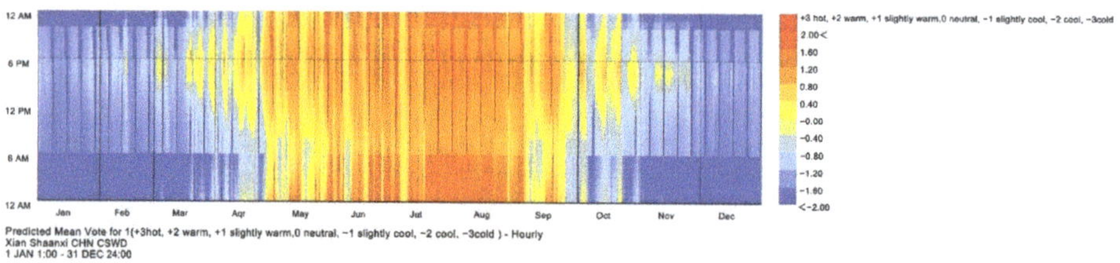

Figure 8. Indoor thermal comfort PMV of CHD Student Center.

According to the simulation results of thermal comfort, the percentage of people with unsatisfactory indoor expectations in the CHD Student Center throughout the year is 23.7%, the absolute value of PMV at this time is 0.89, and the annual indoor discomfort time is 2179.75 h.

The heat transfer coefficient (U value) of each building envelope component is compared with the specified limit value in the building energy-saving design standards across various periods to ascertain compliance with energy-saving design requirements. The comparative results are displayed in Table 6.

Table 6. Judgment of related indicators of CHD Student Center.

Name	Value	Design Standard for Energy Efficiency of Public Buildings GB 50189-2005 Limit Value [37]	Satisfied or Not	Design Standard for Energy Efficiency of Public Buildings GB 50189-2015 Limit Value [42]	Satisfied or Not
Roof $W/(m^2{\cdot}k)$	U = 0.519	$U \leq 0.55$	√	$U \leq 0.45$	×
Exterior wall $W/(m^2{\cdot}k)$	U = 2.055	$U \leq 0.6$	×	$U \leq 0.50$	×

Table 6. *Cont.*

Name		Value	Design Standard for Energy Efficiency of Public Buildings GB 50189-2005 Limit Value [37]	Satisfied or Not	Design Standard for Energy Efficiency of Public Buildings GB 50189-2015 Limit Value [42]	Satisfied or Not
Window wall ratio	East 0.20	U = 3.4	U ≤ 3.0	×	U ≤ 2.7	×
	West 0.20	U = 3.4	U ≤ 3.0	×	U ≤ 2.7	×
	South 0.68	U = 3.4	U ≤ 2.0	×	U ≤ 1.9	×
	North 0.10	U = 3.4	U ≤ 3.5	√	U ≤ 3.0	√
Shape factor		0.15	≤0.40	√	≤0.40	√

Based on the comprehensive findings from field investigations, data testing, and software simulation analyses, the university student center exhibits several issues concerning the thermal performance of its envelope structure, building energy consumption, and indoor thermal comfort:

(1) Thermal performance of the enclosure structure: The table reveals that the CHD Student Center, in its architectural design, does not adhere to building energy conservation requirements. Consequently, the thermal performance of the building envelope exceeds the current codes and standards. There exists substantial potential for energy-saving optimization in the building envelope design.

(2) Building energy consumption: As per the specifications outlined in the "Standard for Energy Consumption of Buildings" GB/T 51161-2016 [43], the prescribed value for Xi'an's annual unit area heat consumption index stands at 0.21GJ/(m^2·a), equivalent to 58.333 kW·h/m^2. However, based on the energy consumption simulation results, the CHD Student Center demonstrates an annual cumulative heat consumption per unit area of 82.15 kW·h/m^2, surpassing Xi'an's building heat consumption index limit.

(3) Building thermal comfort: According to the standard outlined in the "Evaluation Standard for Indoor Thermal Environment in Civil Buildings" GB/T 50785-2012 [39], a thermal environment with a Predicted Percentage of Dissatisfied (PPD) of ≤10% signifies comfort, while the indoor thermal environment should ideally range between −0.5 and +0.5 for Predicted Mean Vote (PMV). Based on the thermal comfort simulation results, the CHD Student Center experiences an unsatisfactory indoor environment throughout the year, with 23.7% of individuals dissatisfied, and a PMV of 0.89, categorized as a grade III standard. This indicates low satisfaction among occupants regarding the humidity and thermal environment. Analysis of the construction drawings reveals that the thermal performance of each component of the building envelope exceeds the current code limit. This situation is detrimental to creating a favorable indoor thermal environment in the building, making it challenging to maintain comfortable conditions.

4. Results and Discussion

Based on the status of student centers in Xi'an and simulation analysis results, it has been discovered that the CHD Student Center has significant potential for energy conservation in terms of both building energy consumption and indoor thermal comfort. Expanding on this assessment and considering the actual influencing factors, a transformative design was implemented. This involved adjusting the external wall insulation materials and thickness, types of external windows, roof insulation materials and thickness, external sunshades, and more. These modifications were evaluated against the heat transfer coefficient specified in the "Design Standard for Energy Efficiency of Public Buildings" GB50189-2015 [42]. Using a multi-objective optimization module, the optimal Pareto solution set was computed to identify the most fitting scheme that meets the heat transfer coefficient prerequisites. The process of multi-objective optimization, along with the resulting optimal design concepts relevant to comparable existing buildings in the Xi'an area, is depicted in Figure 9: schematic diagram of the multi-objective optimization research process and technical platform.

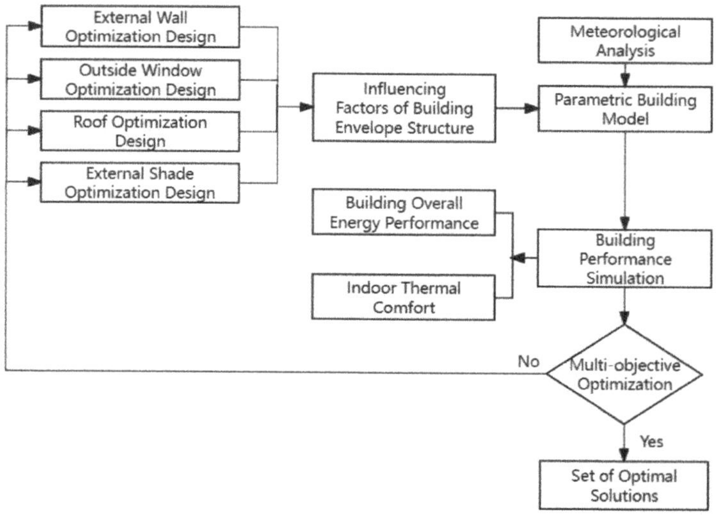

Figure 9. Schematic diagram of multi-objective optimization research process and technology platform of student center.

4.1. Multi-Objective Optimization Experiment Design

The meteorological data selected in this simulation study are the data of typical meteorological years in Xi'an. The schedule of heating and cooling equipment, the opening conditions of ventilation and lighting, and the activity schedule adopted by indoor personnel were consistent with the initial case simulation.

To sum up, the parameter settings of optimization variables in this study are shown in Table 7.

Table 7. Variable parameter setting.

Optimizations	Original Building Parameters	Simulate and Optimize Building Parameters
Roof insulation	35 mm-thick XPS board for roof	R1 100 mm-thick XPS board R2 150 mm-thick XPS board R3 100 mm-thick EPS board R4 150 mm-thick EPS board R5 100 mm-thick rock wool board R6 150 thick rockwool board
Exterior insulation	No external wall insulation	W1 100 mm-thick XPS board W2 150 mm-thick XPS board W3 100 mm-thick EPS board W4 150 mm-thick EPS board W5 100 thick rock wool board W6 150 thick rockwool board
Type of glass for building exterior window	6 mm common double glazing	G1 3 mm + 12Ar + 3 mmLow G2 6 mm + 12Ar + 6 mmLow G3 3 mm + 12Ar + 3 mm plain glass G4 6 mm + 12Ar + 6 mm regular glass
Length of southwest-facing visor	Unshaded	S1 0 m S2 0.5 m S3 1 m S4 1.5 m

Note: Ar in the table indicates filled inert gas.

Referring to the algorithm parameter settings from previous studies [8], a comparative test was conducted regarding the parameter settings in this study. In optimizing the parameter settings of the algorithm, the configuration utilized in this study ensures the algorithm's performance while effectively balancing the optimization of building energy consumption and indoor thermal comfort. The optimization algorithm's parameter settings for this study are detailed in Table 8.

Table 8. Octopus optimization parameter settings for this study.

Name		Value
Elitism	Elitism Ratio	0.50
Mut. Probability	Mutational probability	0.05
Mutation Rate	Mutation rate	0.10
Crossover Rate	Crossover probability	0.80
Population Size	Population size	50
Max Generations	Max evolutionary algebra	50
Record interval	Record time interval	1
Save interval	Save interval	0

4.2. Optimize the Analysis of Experimental Results

In this study, the optimization algorithm module of the software was used to simulate the building energy consumption and the uncomfortable duration inside the building, etc., corresponding to the Pareto scatter plots of 5, 25, 45, and 50 generations (Figures 10–13), until the point distribution reached a stable convergence state. The dark red cube is Pareto's optimal solution set.

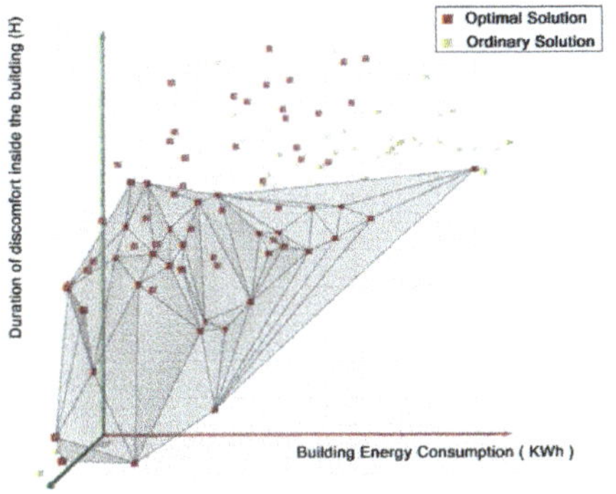

Figure 10. Pareto solution iterated 5 times.

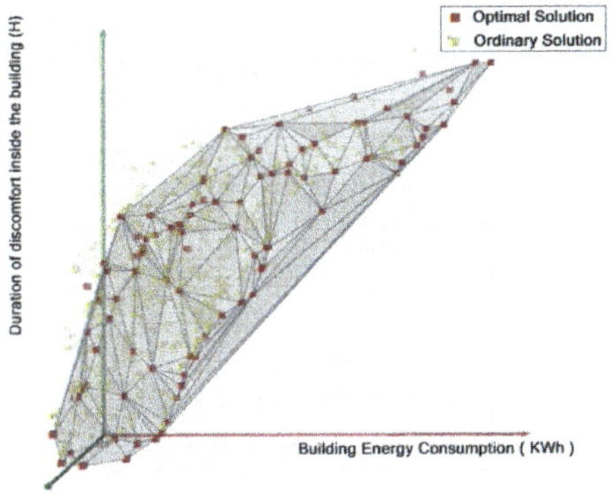

Figure 11. Pareto solution iterated 25 times.

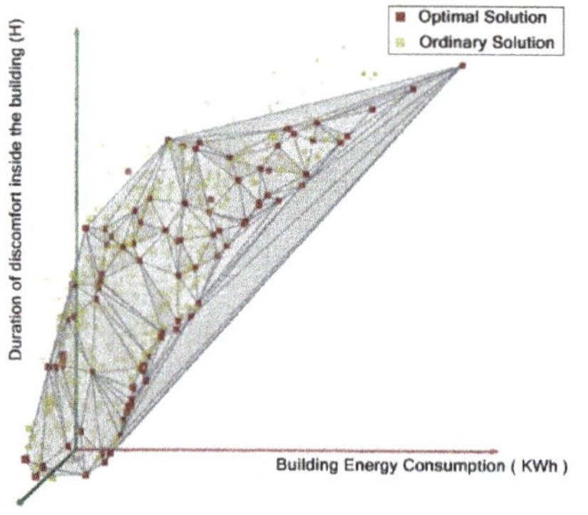

Figure 12. Pareto solution iterated 45 times.

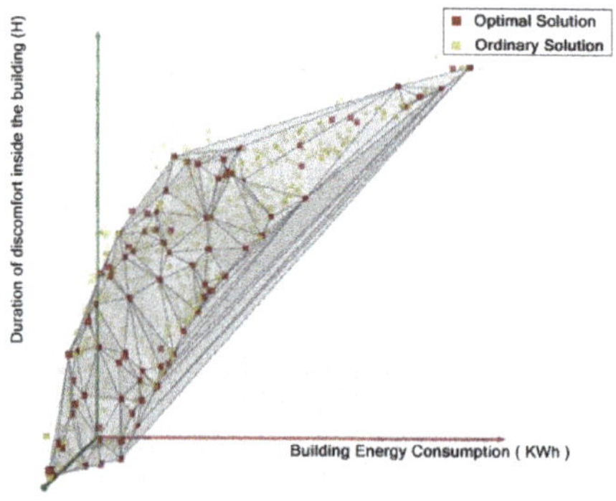

Figure 13. Pareto solution iterated 50 times.

Figures 10–13 illustrate that the Pareto optimal solution sets are interconnected to form a Pareto front surface. The overall shape closely resembles a convex paraboloid centered around the coordinate origin. Observing the fluctuation in the solution set of the 50th-generation frontier surface indicates that this optimization set has converged. The resulting optimal solution set obtained from this convergence can be utilized for further research and analysis. The iterative process of the multi-objective optimization simulation, comprising approximately 2500 operations, unveiled a dense trend in the distribution of the Pareto solution set. The relatively stable position within this distribution can be regarded as the state of convergence for the set.

(1) Analysis of total building energy consumption

Analysis of the 50th-generation Pareto solution set data obtained from the multi-objective optimization study of the CHD Student Center indicates a notable reduction in building energy consumption following the optimization process. The annual energy consumption after optimization ranges between 800,000 kWh and 830,000 kWh. Specifically, the 33rd-generation Pareto solution set (depicted in Figure 14) generated the minimum energy consumption of 800,322.19 kWh through energy consumption optimization simulations. Conversely, the maximum energy consumption was recorded from the 46th-generation Pareto solution set, resulting in 843,523.24 kWh. Overall, in the 50th-generation data set, the building's energy consumption demonstrates a significant reduction compared to the pre-optimization consumption of 1,964,153.07 kWh.

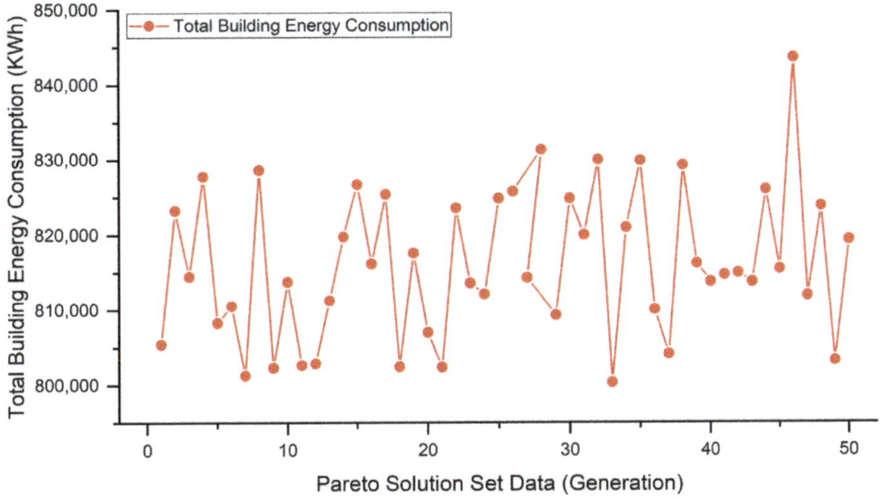

Figure 14. Relationship between building energy consumption and Pareto solution set.

(2) Indoor thermal comfort analysis of buildings

Examining the impact of multi-objective optimization on the discomfort duration within the building, an analysis was conducted using the 50th-generation Pareto solution set data from the multi-objective optimization study of the CHD Student Center. This analysis generated a relationship diagram between the discomfort duration within the building and the Pareto solution set (see Figure 15). The discomfort duration values mostly range between 800 h and 1600 h. As per the multi-objective optimization research, the minimum discomfort duration is recorded in the 8th generation, amounting to 753.52 h. Conversely, the maximum discomfort duration resulting from the multi-objective optimization research occurs in the 21st generation, with a value of approximately 2168.75 h. Overall, across the 50 generations, the building's indoor discomfort duration exhibits a notable decrease compared to the indoor thermal discomfort duration of 2179.75 h before the optimization.

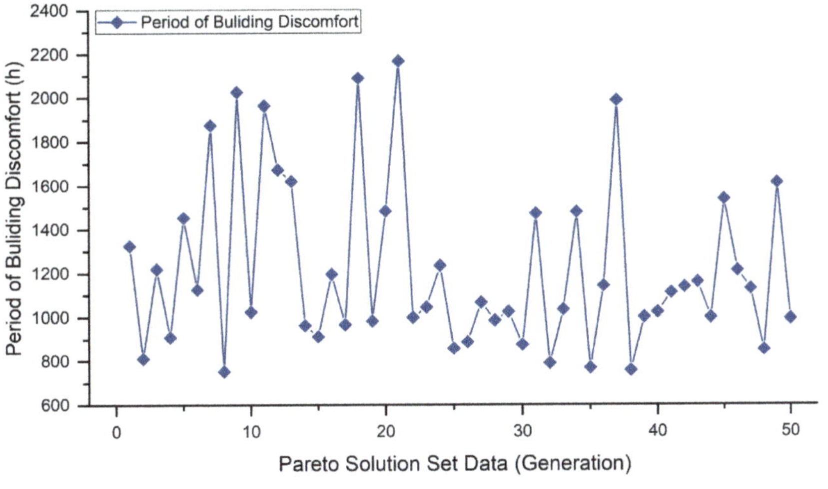

Figure 15. Relationship between indoor discomfort duration and Pareto solution set.

(3) Analysis of variable optimization design

Optimization of wall insulation structure: An analysis of the 50th-generation Pareto solution set reveals that the optimization primarily centers on two types of wall insulation structures: W2 (XPS, 150 mm thickness) and W4 (EPS, 150 mm thickness). Within these options, the XPS board configuration dominates the solution set, suggesting its superior performance in terms of energy consumption and comfort at a thickness of 150 mm.

Optimization of roof insulation structure: An analysis of the Pareto solution set highlights the preference for solutions centered around R1, R2, and R4 (involving XPS and EPS insulation panels with 100 mm and 150 mm thicknesses, respectively). Due to the relatively small roof insulation area, the optimization process does not significantly increase the thickness of the insulation material. This indicates that both XPS and EPS exhibit effective insulation properties, with 150 mm thickness being the prevalent choice.

Selection of external window glass type: The results from multi-objective optimization indicate that G1 (3 mm + 12 Ar + 3 mmLow glass) and G3 (3 mm + 12 Ar + 3 mm ordinary glass) window types are preferred. In particular, the G1 type holds a notably higher proportion within the solution concentration, highlighting its distinct advantages in optimization.

Optimization of horizontal shading: Data analysis shows that S2 (0.5 m overhang horizontal shading) in the southwest direction contributes more effectively to enhancing the energy efficiency and comfort of the building interior. Pareto solution sets without horizontal sunshade (S1) also exist, but their proportion is small.

(4) Weighing scheme analysis

In this study, the scheme provided by the 29th-generation Pareto solution set has been chosen as the optimization scheme (see Figure 16). Based on the multi-objective optimization point diagram of the building, the thermal insulation structure of the external wall has been adjusted to a 150 mm-thick XPS board from the original structural design. The roof's insulation structure has also been set to a 150 mm-thick XPS board. Additionally, the external window glass has been replaced with a 3 mm + 12 Ar + 3 mmLowE structure, and a 0.5 m horizontal sunshade has been installed on the building's southwest side. Simulation results demonstrate a notable reduction in the building's energy consumption, bringing it down to 809,273.24 kWh—considerably lower than the pre-optimization consumption of 1,964,153.07 kWh. Moreover, the discomfort duration inside the building has decreased to 1024.19 h. The corresponding parameters of this solution set are detailed in Table 9.

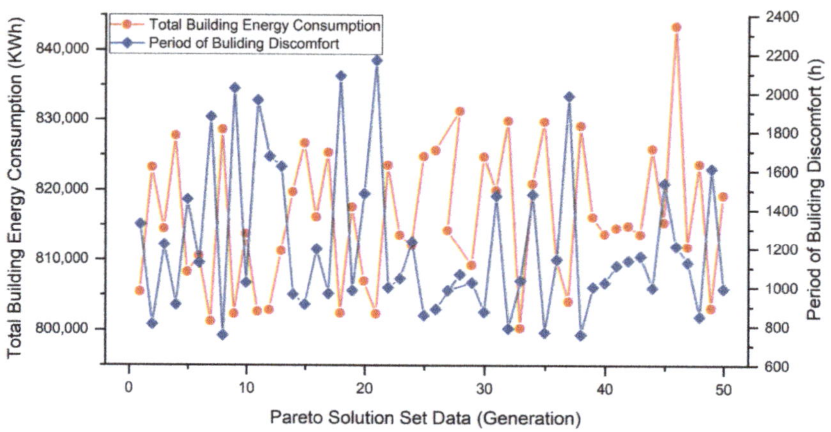

Figure 16. Pareto solution set of building energy consumption and indoor discomfort duration.

In the research process, if there is an optimal solution or solution set scheme, will compare the benchmark values in the building energy efficiency design codes or building comfort standards, including the Evaluation Standard for Indoor Thermal Environment in Civil Buildings GB/T 50785-2012 and the Design Standard for Energy Efficiency of Public Buildings

GB50189-2015. When there is no optimal solution set, it is necessary to calculate the data of the non-optimal solution set calculated in Ladybug and Honeybee software Version 0.0.65, calculate the heat transfer coefficients of the corresponding parts of the model, such as the exterior windows, exterior walls, and roof, and then compare them with the benchmark heat transfer coefficients in the standard. Based on this, select the optimal solution that meets the heat transfer coefficient requirements. Based on this, compare the energy consumption and uncomfortable time length of the selected solution to determine the optimal solution.

Table 9. Corresponding parameters of the solution set of the 29th-generation Pareto.

Algebra	Building Energy Consumption (KWh)	Duration of Discomfort inside the Building (h)	Type of Exterior Insulation	Type of Roof Insulation	Exterior Window Glazing Type	Southwest Horizontal Shade (m)
Pareto 29	809,273.24	1024.19	W2	R2	G1	S2

4.3. Multi-Objective Optimization of Optimal Solution Set

Calculate the energy-saving rate of the envelope structure according to the "Standard for Green Performance Calculation of Civil Buildings" JGJ/T 449-2018 [44]. The calculation formula is as follows (Equation (1)):

$$\eta_E = \frac{E_B - E_0}{E_0} \times 100\% \quad (1)$$

In the formula, the following are defined:

η_E—total building optimization rate, %;
E_B—the total building comprehensive energy consumption after optimization, kWh;
E_0—total building comprehensive energy consumption after initial building model, kWh.

After evaluating the building's energy consumption and thermal comfort performance target optimization rate, we compared the multi-objective optimization data with the original building data. The annual comprehensive energy consumption was reduced by approximately 58.8% following optimization. Simultaneously, the annual indoor thermal comfort duration increased by about 53.0%. This validates and analyzes the energy-saving optimization scheme for the CHD Student Center as an effective transformation strategy, detailed in Figures 17 and 18, and Table 10.

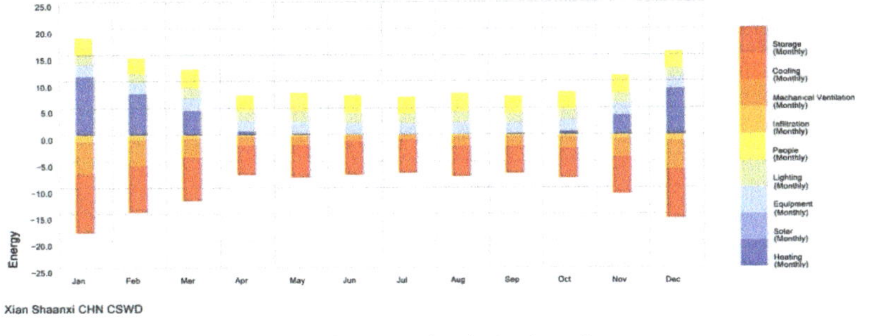

Figure 17. Annual energy balance diagram of optimization scheme.

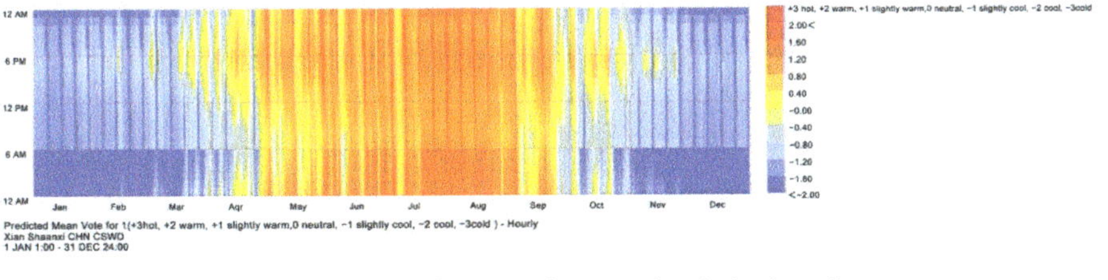

Figure 18. Annual indoor thermal comfort PMV diagram of optimization scheme.

Table 10. Corresponding parameters of the initial scheme and the solution set after optimization.

Name	Building Energy Consumption (KWh)	Duration of Discomfort inside the Building (h)	Type of Exterior Insulation	Type of Roof Insulation	Exterior Window Glazing Type	Southwest Horizontal Shade (m)
Original	1,964,153.07	2179.75	/	35 mm-thick XPS board	6 mm plain double glazing	/
Optimized Pareto29 scheme	809,273.24	1024.19	150 mm-thick XPS insulation board	150 mm-thick XPS insulation board	3 mm + 12Ar + 3 mmLow	The southwest-facing horizontal visor protruding length is 0.5 m

After conducting a comparative analysis of the annual comprehensive energy consumption and indoor thermal comfort values before and after the optimization plan, it is concluded that the case building's annual comprehensive energy consumption was reduced by 58.8% after implementing optimization measures, now being 75.36 $kW{\cdot}h/m^2$, much lower than the 183.57 $kW{\cdot}h/m^2$ before the renovation; the heat consumption index is 38.19 $kW{\cdot}h/m^2$, reduced by 46.5%, which meets the limit value of 58.333 $kW{\cdot}h/m^2$ for the building heat consumption index in Xi'an according to the "Standard for Energy Consumption of Buildings" GB/T 51161-2016. The indoor thermal comfort duration has increased by 53.0%, and the percentage of indoor dissatisfaction expected throughout the year is 11.2%. According to the "Evaluation Standard for Indoor Thermal Environment in Civil Buildings" GB/T 50785-2012, it is classified as Level II standard and basically meets the comfort requirements. These verification results demonstrate the rationale and feasibility of the optimization process.

Based on the above analysis, it can be seen that unlike single-factor optimization, multi-objective optimization can be optimized from both energy consumption and thermal comfort perspectives. For various practical cases, it is necessary to reasonably adjust the various influencing factors involved in multi-objective optimization based on the actual situation of the case, and ensure the optimal optimization plan through repeated adjustments. Not only that, other variables can be added during the optimization process and multi-objective joint optimization design can be carried out to obtain the Pareto solution set through simulation calculation. In addition, targeted economic strategies can also be proposed based on analysis of the actual situation, such as adjusting material thickness reasonably, selecting appropriate material categories, etc., thus forming the optimal plan to achieve the maximum economic benefits.

5. Conclusions

This study focuses on the university student center buildings in China and proposes a multi-objective optimization design method for achieving thermal comfort while maintaining low energy consumption. The study illustrates this method through practical examples, leading to the following conclusions:

(1) The university student center serves diverse functions, encompasses various spaces, exhibits high total energy consumption, and displays energy usage patterns distinct from other university buildings. It demonstrates significant potential for energy savings.

(2) The Octopus multi-objective optimization module is employed to conduct research, facilitating the simultaneous optimization of building comprehensive energy consumption and indoor thermal comfort. This approach allows for comprehensive design optimization involving elements such as the building's exterior wall (insulation material type and thickness), exterior windows (glass material type and thickness), roof (insulation material type and thickness), and external sunshade (length of horizontal sunshade). Initially, this scheme aids architects in making informed design decisions during the project's early stages, fostering energy-conserving building design. Furthermore, it generates more practical transformation technical schemes. In this study, the annual comprehensive energy consumption of the

building is reduced by 58.8%, while the indoor thermal comfort duration increases by 53.0%, demonstrating the feasibility and soundness of the optimization design method.

(3) The multi-objective optimization platform considers multiple factors globally, utilizing a multi-objective optimization algorithm to prevent insufficient consideration of individual factors. Moreover, this process generates multiple visualized optimization schemes, allowing designers to select options based on different objectives. This ensures that all objective functions achieve an optimal state, significantly enhancing the efficiency and effectiveness of building transformation optimization.

In the optimization of university student center design, this study focused solely on analyzing building energy consumption and indoor thermal comfort, while other variables were not extensively compared or discussed. Future optimization experiments could benefit from employing multi-objective optimization algorithms to consider comprehensive factors such as site selection, space layout, and building equipment selection, enabling a more holistic examination of the university student center. Additionally, conducting more in-depth analyses and comparisons of building performance across various climate zones in subsequent research, through detailed data or statistical analysis, would facilitate broader conclusions.

Author Contributions: Conceptualization, M.L. and Q.L.; methodology, M.L., Y.Q. and Q.L; software, C.Y.; validation, Y.Q., N.Y. and C.Y.; formal analysis, M.L. and Q.L.; investigation, N.Y. and C.Y.; resources, Y.Q.; data curation, N.Y.; writing—original draft preparation, M.L., Y.Q., N.Y. and Q.L.; writing—review and editing, M.L. and Q.L.; visualization, Y.Q. and N.Y.; supervision, Q.L.; project administration, M.L. and Q.L.; funding acquisition, M.L. and Q.L. All authors have read and agreed to the published version of the manuscript.

Funding: This research is funded by Shaanxi Province Housing and Urban Rural Construction Science and Technology Plan Project of Shaanxi Provincial Department of Housing and Urban Rural Development, funding number: 2021-K29; it is also funded by Special Fund Project for Basic Research Business Expenses of Central Universities of Chang'an University, funding number: 300102412605.

Data Availability Statement: The data presented in this study are available on request from the corresponding author.

Conflicts of Interest: Author Chongyi Yan was employed by CCCC First Highway Consultant Co., Ltd. The remaining authors declare that the research was conducted in the absence of any commercial or financial relationships that could be construed as a potential conflict of interest.

References

1. Zhuang, W.; Liu, J.; Wang, J.; Mei, H.; Ji, J.; Lin, B.; Miao, Z.; Geng, Y. Key Frontier basic science issues in building carbon neutrality. *Sci. Found. China* **2023**, *37*, 348–352.
2. Caldas, L.G.; Norford, L.K. A design optimization tool based on a genetic algorithm. *Autom. Constr.* **2002**, *11*, 173–184. [CrossRef]
3. Magnier, L.; Haghighat, F. Multi objective optimization of building design using TRNSYS simulations, genetic algorithm, and Artificial Neural Network. *Build. Environ.* **2010**, *45*, 739–746. [CrossRef]
4. Fang, Y.; Cho, S. Design optimization of building geometry and fenestration for daylighting and energy performance. *Sol. Energy* **2019**, *191*, 7–18. [CrossRef]
5. Li, H. *Research on Optimization Method of Building Shape Coefficient and Upwind Area Ratio Based on Genetic Algorithm*; South China University of Technology: Guangzhou, China, 2015.
6. Kiss, B.; Silvestre, J.D.; Madeira, J.F.A.; Santos, R.A.; Szalay, Z. Environmental and economic optimization of buildings for different climates. *IOP Conf. Ser. Earth Environ. Sci.* **2020**, *588*, 032033. [CrossRef]
7. Giouri, E.D.; Tenpierik, M.; Turrin, M. Zero energy potential of a high-rise office building in a Mediterranean climate: Using multi-objective optimization to understand the impact of design decisions towards zero-energy high-rise buildings. *Energy Build.* **2020**, *209*, 109666. [CrossRef]
8. Chegari, B.; Tabaa, M.; Simeu, E.; Moutaouakkil, F.; Medromi, H. An optimal surrogate-model-based approach to support comfortable and nearly zero energy buildings design. *Energy* **2022**, *248*, 123584. [CrossRef]
9. Bre, F.; Roman, N.; Fachinotti, V.D. An efficient metamodel-based method to carry out multi-objective building performance optimizations. *Energy Build.* **2020**, *206*, 109576. [CrossRef]
10. Albatayneh, A. Optimizing the parameters of a building envelope in the East Mediterranean Saharan, cool climate zone. *Buildings* **2021**, *11*, 43. [CrossRef]
11. Asadi, E.; Da Silva, M.G.; Antunes, C.H.; Dias, L. A multi-objective optimization model for building retrofit strategies using TRNSYS simulations, GenOpt and MATLAB. *Build. Environ.* **2012**, *56*, 370–378. [CrossRef]

12. Erlendsson, Ö. *Daylight Optimization—A Parametric Study of Atrium Design: Early Stage Design Guidelines of Atria for Optimization of Daylight Autonomy*; Royal Institute of Technology Stockholm: Stockholm, Sweden, 2014.
13. Belloni, E.; Buratti, C.; Merli, F.; Moretti, E.; Ihara, T. Thermal-energy and lighting performance of aerogel glazings with hollow silica: Field experimental study and dynamic simulations. *Energy Build.* **2021**, *243*, 110999. [CrossRef]
14. Kiss, B.; Szalay, Z. Modular approach to multi-objective environmental optimization of buildings. *Automat. Constr.* **2020**, *111*, 103044. [CrossRef]
15. Zhang, A. Research on Multi-Objective Optimization Design of Primary and Secondary Schools in Cold Regions Based on Energy Consumption and Comfort. Ph.D. Thesis, Tianjin University, Tianjin, China, 2018.
16. Zhao, Y.; Yang, J.; Fang, Z.; Zhang, X.; Guo, T.; Li, Y. Passive design strategies to improve student thermal comfort: A field study in semi-outdoor spaces of academic buildings in hot-humid areas. *Urban Clim.* **2024**, *53*, 101807. [CrossRef]
17. Attaianese, E.; d'Ambrosio Alfano, F.R.; Palella, B.I.; Pepe, D.; Vanacore, R. An integrated methodology of subjective investigation for a sustainable indoor built environment. The case study of a university campus in Italy. *Atmosphere* **2021**, *12*, 1272. [CrossRef]
18. Liu, Q. Research on High Dimensional Multi-Object Optimization and Decision Support Method of Architecture in Scheme Design Stage. Master's Thesis, Harbin Institute of Technology, Harbin, China, 2020.
19. He, L.; Zhang, L. A bi-objective optimization of energy consumption and investment cost for public building envelope design based on the ε-constraint method. *Energy Build.* **2022**, *266*, 112133. [CrossRef]
20. Nasrollahzadeh, N. Comprehensive building envelope optimization: Improving energy, daylight, and thermal comfort performance of the dwelling unit. *J. Build. Eng.* **2021**, *44*, 103418. [CrossRef]
21. Zhao, D.; McCoy, A.P.; Du, J.; Agee, P.; Lu, Y. Interaction effects of building technology and resident behavior on energy consumption in residential buildings. *Energy Build.* **2017**, *134*, 223–233. [CrossRef]
22. Siddhartha, M.Y.P.; Pai, Y. Effect of building orientation and window glazing on the energy consumption of HVAC system of an office building for different climate zones. *Int. J. Eng. Res.* **2015**, *4*, 838–843. [CrossRef]
23. Ruiz, G.R.; Bandera, C.F.; Temes, T.G.A.; Gutierrez, A.S.O. Genetic algorithm for building envelope calibration. *Appl. Energy* **2016**, *168*, 691–705. [CrossRef]
24. Wu, T.; Wang, B.; Zhang, D.; Zhao, Z.; Zhu, H. Benchmarking Evaluation of Building Energy Consumption Based on Data Mining. *Sustainability* **2023**, *15*, 5211. [CrossRef]
25. Fesanghary, M.; Asadi, S.; Geem, Z.W. Design of low-emission and energy-efficient residential buildings using a multi-objective optimization algorithm. *Build. Environ.* **2012**, *49*, 245–250. [CrossRef]
26. Elbeltagi, E.; Wefki, H.; Khallaf, R. Sustainable Building Optimization Model for Early-Stage Design. *Buildings* **2022**, *13*, 74. [CrossRef]
27. Liu, B.; Pouramini, S. Multi-objective optimization for thermal comfort enhancement and greenhouse gas emission reduction in residential buildings applying retrofitting measures by an Enhanced Water Strider Optimization Algorithm: A case study. *Energy Rep.* **2021**, *7*, 1915–1929. [CrossRef]
28. Ziaee, N.; Vakilinezhad, R. Multi-objective optimization of daylight performance and thermal comfort in classrooms with light-shelves: Case studies in Tehran and Sari, Iran. *Energy Build.* **2022**, *254*, 111590. [CrossRef]
29. Zhang, Z.; Wang, W.; Song, J.; Wang, Z.; Wang, W. Multi-Objective Optimization of Ultra-Low Energy Consumption Buildings in Severely Cold Regions Considering Life Cycle Performance. *Sustainability* **2022**, *14*, 16440. [CrossRef]
30. Taylor, M.; Brown, N.C.; Rim, D. Optimizing thermal comfort and energy use for learning environments. *Energy Build.* **2021**, *248*, 111181. [CrossRef]
31. Lakhdari, K.; Sriti, L.; Painter, B. Parametric optimization of daylight, thermal and energy performance of middle school classrooms, case of hot and dry regions. *Build. Environ.* **2021**, *204*, 108173. [CrossRef]
32. Hosamo, H.H.; Tingstveit, M.S.; Nielsen, H.K.; Svennevig, P.R.; Svidt, K. Multi objective optimization of building energy consumption and thermal comfort based on integrated BIM framework with machine learning-NSGA II. *Energy Build.* **2022**, *277*, 112479. [CrossRef]
33. Lin, C.J.; Wang, K.J.; Dagne, T.B.; Woldegiorgis, B.H. Balancing thermal comfort and energy conservation–A multi-objective optimization model for controlling air-condition and mechanical ventilation systems. *Build. Environ.* **2022**, *219*, 109237. [CrossRef]
34. Lin, Y.; Zhao, L.; Liu, X.; Yang, W.; Hao, X.; Tian, L. Design optimization of a passive building with green roof through machine learning and group intelligent algorithm. *Buildings* **2021**, *11*, 192. [CrossRef]
35. *Architectural Design Data Collection (Volume 3)*; China Architecture & Building Press: Beijing, China, 2017.
36. *JGJ/T 41-2015*; Code for the Architectural Design of Cultural Centers. China Architecture & Building Press: Beijing, China, 2015.
37. *GB50189-2005*; Design Standard for Energy Efficiency of Public Buildings. China Architecture & Building Press: Beijing, China, 2005.
38. *GB 50176-2016*; Code for Thermal Design of Civil Buildings. China Architecture & Building Press: Beijing, China, 2016.
39. *GB/T 50785-2012*; Evaluation Standard for Indoor Thermal Environment in Civil Buildings. China Architecture & Building Press: Beijing, China, 2012.
40. Li, C.; Chen, Y. A multi-factor optimization method based on thermal comfort for building energy performance with natural ventilation. *Energy Build.* **2023**, *285*, 112893. [CrossRef]
41. *GB/T1883-2022*; Standards for Indoor Air Quality. State Administration for Market Regulation: Beijing, China, 2022.
42. *GB50189-2015*; Design Standard for Energy Efficiency of Public Buildings. China Architecture & Building Press: Beijing, China, 2015.

43. *GB/T 51161-2016;* Standard for Energy Consumption of Building. China Architecture & Building Press: Beijing, China, 2016.
44. *JGJ/T 449-2018;* Standard for Green Performance Calculation of Civil Buildings. China Architecture & Building Press: Beijing, China, 2018.

energies

MDPI

Article

Optimization of Renewable-Based Multi-Energy Systems in Residential Building Design

Vasileios Kilis [1], Georgios Anastasiadis [1], Nikolaos Ploskas [2] and Giorgos Panaras [1,*]

[1] Mechanical Engineering Department, University of Western Macedonia, 50131 Kozani, Greece; v.kilis@uowm.gr (V.K.); gdanastasiadis@gmail.com (G.A.)
[2] Electrical and Computer Engineering Department, University of Western Macedonia, 50131 Kozani, Greece; nploskas@uowm.gr
* Correspondence: gpanaras@uowm.gr

Abstract: Electrification is a key priority of the European Union, focusing on saving energy resources and mitigating carbon emissions through enhancing restrictions on relative policies and initiatives. For such goals to be achieved, investing in renewable energy technologies on large- and small-scale projects is promoted. These efforts were implemented in the building sector too, highlighting the importance of optimal decisions in improving the energy performance of buildings, from an economic, energy and environmental perspective. In this context, this paper aims to elaborate a decision-making methodology for building thermal design, considering the optimal selection and operation of multi-energy systems focused on renewable technologies. Solar thermal collectors, photovoltaic systems and heat pumps were included in an Energy Hub for meeting the heating, cooling and domestic hot water energy demand. Optimal decisions were achieved by formulating Mathematical Programming models in GAMS, for minimizing economic, energy and environmental parameters of the systems under a life cycle perspective. The proposed methodology was implemented in a residential building case study. Results show that combining heat pumps with photovoltaics is preferable for all of the examined criteria, while a sensitivity analysis of the economic, energy and environmental parameters, influencing the energy mixture, leads to optimal solutions with the participation of different energy systems.

Keywords: building energy efficiency; optimal decision making; multi-energy systems; multiple criteria optimization; energy hub; mathematical programming

1. Introduction

Social development and rising economies worldwide are considered as key motivational aspects for governments to ensure independence on covering energy requirements. Additional concerns, like the increase in energy demand, rising oil prices and fossil fuel depletion, highlight the need for energy security under the perspective of affordable and sustainable energy pricing and access. This is indicative for the electricity supplies and other energy-intensive uses, as the electricity sector seems to present a rising growth in global energy consumption, increasing its share of penetration from 15% in 2000 to around 19% in 2017 [1]. In 2022, in the European Union (EU), the share of electricity in total energy consumption was 23%, second to the rating of all energy products. All above are in line with the energy goals of mitigating greenhouse gas (GHG) emissions and exploiting technologies utilizing renewable energy sources (RESs). This is characteristic in EU 2022, as renewable energy contributes to the highest share (43%) in energy consumption [2].

Among all energy consumers (industry, transport, agriculture, etc.), the building sector represents a significant portion of nearly one-third of global energy use, and a great contribution to GHG emissions. In fact, building operations account for approximately 55% of global electricity demand, although the goals of the Paris Agreement propose a decrease by 45% by 2030 of the building energy usage, which necessitates an annual rate that is five times greater than the progress made in recent years [3,4]. Furthermore, the building stock in the EU is responsible for about 40% of total energy consumption [5,6], and accounts for about 36% of total GHG emissions too [7]. This is due to estimations mentioning that almost 75% of existing buildings are considered as energy inefficient because of their high energy consumption for heating, cooling, lighting, etc. [8]. This value was higher (97% of EU buildings) when the decarbonization goals of 2050 were considered [9], even if the contribution of the residential sector towards emission reduction targets, with a 34% reduction between 2005 and 2022, was achieved [10]. Also, the current rate of deep energy renovation for achieving primary energy savings (beyond 60%) and high energy performance in buildings stands at 0.2%, which is far away from the goals of achieving 55% GHG emission reduction (compared to 1990) by 2030 and carbon neutrality by 2050 [11]. Also, deep building energy renovations, which account for 70% of the total, should increase to a rate of 3% [12].

The EU has developed a comprehensive series of initiatives and directives, considering the improvement of building energy efficiency and decarbonization, progressing toward climate neutrality by 2050. In this context, the Clean Energy Package was introduced in 2018, focusing on accelerating the energy transition in Europe by improving building energy efficiency and integrating RES, as well as setting requirements for new constructions to meet nearly zero-energy standards [13]. This package includes relevant directives and regulations, such as the Energy Efficiency Directive (EED), mentioning the reduction of energy demand through energy savings targets, and the Energy Performance of Buildings Directive (EPBD), focusing on improving building energy efficiency through retrofitting and energy renovations, while outlining a comprehensive vision of the complete electrification of residential buildings' energy uses, as depicted in its recent recast [14]. Also, the Renewable Energy Directive (RED) promotes the integration of RES, and the Governance Regulation (GOV) establishes frameworks for national energy and climate plans (NECPs) to monitor progress toward EU targets (especially for 2030).

In the same concept, the COVID-Related Recovery Plan for Europe provided funds for sustainable recovery. This plan includes several programs focusing on building energy renovation through initiatives, such as the EU Green Deal, the Renovation Wave and the "Fit for 55" package, setting strategic priorities for the 2019–2024 period [13]. More specifically, the EU Green Deal provides an extensive roadmap for GHG neutrality by 2050 considering several sectors (economy, environment, energy, mobility, agriculture, etc.), aiming to conserve, while also enhancing, the EU natural capital, and protecting citizens from environment-related risks and impacts too. It is recognized that the building sector is important for the decarbonization effort, setting goals such as doubling energy renovation rates, mitigating energy poverty, and generating green jobs [15]. This led to the "Fit for 55" package being released in 2021, which includes revisions of the EPBD, EED and RED, essentially aligning them with the targets of reducing GHG emission by 55% by 2030. The main provisions included in this package consider the transition from nearly zero-energy buildings (NZEBs) to zero-emission buildings (ZEBs), the introduction of Building Renovation Passports, the definition of lifecycle GHG emissions and the identification of Minimum Energy Performance Standards (MEPSs) for significant building retrofitting. In addition, the Renovation Wave initiative, which was launched alongside "Fit for 55", was set under the perspective of the Green Deal [16]. This initiative focuses on the decarbonization of the building sector, emphasizing heating and cooling systems for public and residential buildings and addressing energy poverty too. It also provided financial and technical assistance for strengthening the incentives for large-scale renovations. The REPowerEU Plan, introduced in 2022, enhanced the beforementioned goals by reducing dependence on fossil fuels (including natural gas) and scaling up the deployment of RES and electrification, considering the energy crisis caused by geopolitical tensions [17].

Considering the energy problem in the building sector, as well as the goals and measures set into several initiatives, the necessity for implementing advanced energy-management strategies and technologies is highlighted. Interventions in building envelopes via the implementation of different energy-saving measures (ESMs), such as enhancing thermal insulation, as well as the decision-making for the energy systems (energy supply systems, ESSs) to meet energy demand (heating, cooling, air conditioning and electricity) are considered crucial parameters for improving building energy efficiency. The decisions made for ESMs and ESSs are essential for the development of efficient energy-management strategies, influencing building sustainability and performance throughout its entire life-cycle, especially when decisions were made in the early design stages. In this context, considerable efforts have been devoted to improving building energy efficiency through different strategies by examining several factors, such as indoor and outdoor climate, building fabric, energy systems, the behaviour of the occupants, etc., under the perspective of economic, energy, environmental, sustainable, technical, social and many other aspects. Such examples include the integration of renewable-based technologies into buildings [18], the use of high-performance energy systems and devices [19], the development of innovative design methods [20], like multi-energy systems and Energy Hubs (EHs) [21,22], the creation of appropriate mathematical models describing aspects from building operation [23], the assurance of indoor thermal comfort through energy-efficient solutions [24], the improvement of thermal insulation in building envelopes [25], and more.

All of the above formulate a multi-parameter decision-making problem for building design, where different alternatives should be examined under an optimization process [26]. Such processes can be described as an attempt to find the optimal values for a set of variables (decision variables), while satisfying various constraints, under the optimization of specific objectives. Different optimization methods can be combined with several dynamic simulation tools such as EnergyPlus 24.2.0 and TRNSYS® 18.05.0001, in order to manage complex design spaces and identify optimal solutions [27–29]. The development of multi-energy systems, combined with the concept of EHs, provides a solution for implementing optimization methodologies in building thermal design. EHs set the coupling of different energy carriers, under the perspective of importing, producing, converting and storing different types of energy, while its respective optimization process proposes optimal decisions considering the selection of different energy systems.

The growing environmental and energy concerns mentioned above highlight the need for adopting the concept of multi-energy systems, combining multiple energy sources (conventional and renewable) with different energy uses. This is indicative in the literature, where the deployment of integrated electricity and heating systems seems to be a strategic response. For example, in [30], a novel method for providing an optimal time-scaling matching for integrated electricity and heating systems was developed, while in [31], the integration of gas-grid models into simulation tools for capturing their operation was depicted. Furthermore, the concept of multi-energy systems and their coupling with the EHs in optimization problems for the building sector is commonly used, especially under modelling, operation and planning issues. In [32], a literature review on the status of multi-energy systems, together with evaluation methods for the penetration of RESs in the building sector is presented. Moreover, ref. [33] modelled a multi-energy system to meet the thermal and electrical needs of 40 residential buildings, utilizing solar and wind energy, as well as a cogeneration unit. Furthermore, ref. [21] developed an EH with multi-energy systems for optimizing the design of a residential building. The systems examined were photovoltaic systems, a gas condensing boiler, heat pumps, photovoltaic thermal collectors and electric chillers, with the aim of minimizing investment costs and the use of non-renewable sources. A novel strategy for residential and commercial EH optimization was developed in [34] by integrating specific demand response programs and time-of-use tariffs for achieving cost minimization and improvement in resource scheduling considering uncertain conditions. Similarly, ref. [35] proposed a multi-objective optimization control strategy for EHs within multi-energy systems. They aimed at distributing thermal and electrical loads in an efficient manner, opting for balanced solutions between economic and environmental criteria. Also, ref. [36] proposes an energy management strategy for the operation of integrated electricity and heating systems in intelligent buildings (the interconnection of photovoltaic and heat pumps), for meeting conventional electrical and heating loads of buildings. The objective was the minimization of the overall system cost, considering alternative scheduling scenarios for energy supply and demand.

The literature mentioned above highlights the need to develop reliable and flexible decision-making methodologies for improving the process of building thermal design and enhancing their efficiency. Also, the concept of connecting EHs with multi-energy systems under optimization processes is used in many studies, where the use of integrated electricity and heating systems is highlighted, especially for defining management strategies for their operation. Many studies focus on obtaining cost-effective operation schemes, considering the energy production side under optimal multi-energy coordination [37,38]. Net-metering is a simpler approach for energy management, utilized in this study. More specifically, net-metering is a regulatory financial mechanism designed to promote the adoption of photovoltaic systems by enabling a bidirectional flow of electricity between the grid and consumers, under energy management. The surplus electricity produced by photovoltaics is fed back into the utility grid, while, when such production is limited, the electricity demand can be met by the grid. The net-metering strategy is based on balancing electricity production and consumption, reducing the dependency on the grid.

Although most of such studies can integrate dynamic simulations, their goal is limited to one criterion, like the minimization of the overall economic costs. However, the proposed methodology implements a decision-making process, considering several criteria, giving a life cycle perspective too. More specifically, this study is focused on renewable-based multi-energy systems, following the EU goal of increasing the use of RESs. In this context, heat pumps, combined with photovoltaic systems or the electricity grid and solar thermal collectors, were examined, giving a life cycle perspective on their construction and operation, which is an added value for such studies. In addition to this, the present analysis incorporates a novel model utilizing the f-chart method for sizing the solar thermal collectors. The main scope of this study is to provide optimal decisions for the sizing and operation of systems, in order to meet the thermal energy demand of a building (space heating and cooling and domestic hot water—DHW). The objectives of the optimization include multiple conflicting criteria, considering economic, energy and environmental parameters under a life cycle perspective (embodied energy and GHG emissions), bolstering the innovativeness of this study. The optimization process was based on the development of Mathematical Programming (MP) models in the General Algebraic Modelling System (GAMS), and the SCIP solver was used for providing the optimal solutions. The proposed methodology was implemented to a case study multi-storey residential building. A sensitivity analysis was conducted considering economic and environmental parameters, as well as different primary energy factors, depending on alternative energy mixtures constituting the electricity grid, performing an innovative investigation. Finally, the optimal decisions resulting from the GAMS were compared with the ones provided by a brute force analysis.

2. Materials and Methods

2.1. Framework of the Decision-Making Process

The overall framework of the proposed decision-making methodology considers several essential steps, as depicted in Figure 1, in order for the optimal solutions to be provided, upon the concept of improving economic, energy and environmental aspects of building energy systems. The first key parameters include identifying the geometric and thermal properties of the building envelope. The utilization of an energy simulation tool, considering the beforementioned characteristics of the building and the climatic conditions, are essential for calculating the energy demand for heating, cooling, DHW, etc. Afterwards, the architecture of the EH should be formulated, including the key parameters characterizing the thermal behaviour of the building envelope, as well as the basic aspects of the multi-energy systems involved. The concept of the EH incorporates an optimization methodology, considering alternative criteria for minimizing economic, energy and environmental aspects. The arising optimization problem was approached by developing MP models in GAMS, in order for the optimal decisions to be defined.

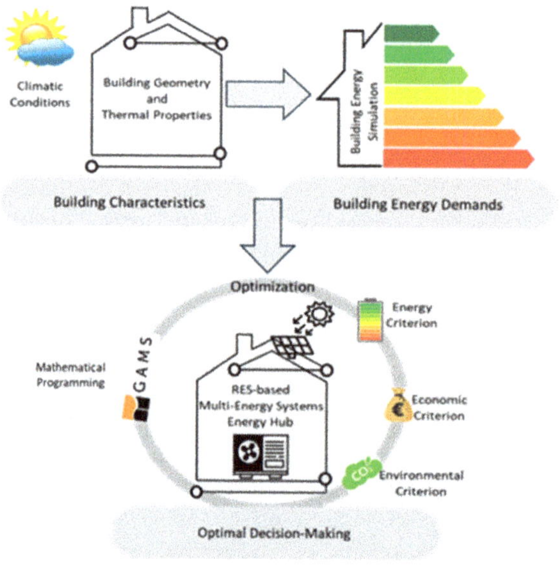

Figure 1. Framework of the proposed decision-making methodology.

2.2. Building the Case Study

The case study building is a 4-storey residential building, with a total floor plan surface area of about 390 m^2 (Figure 2). The basic geometric characteristics are presented in Table 1. The building was constructed in Thessaloniki, Greece, which is considered to belong to climatic zone C, according to the Greek EPBD for climatic zone C [39]. The U-values of the basic building envelope components are presented in Table 2, leading to a well-insulated building (Um = 0.74 W/m^2K).

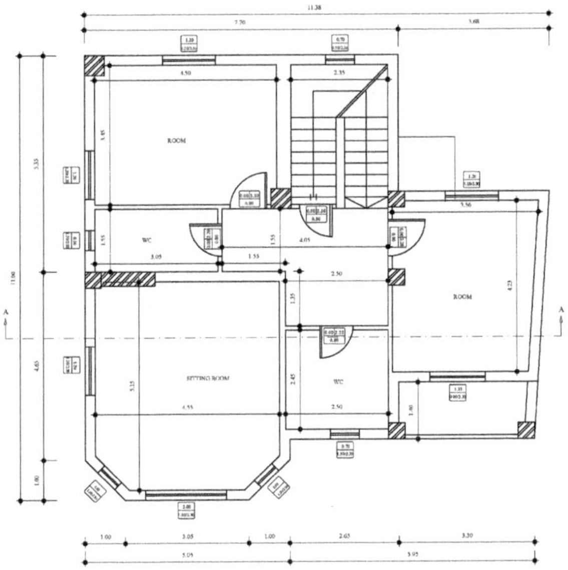

Figure 2. Floor plan of a typical storey of the case study building.

Table 1. Geometric parameters of the case study building envelope.

Geometric Parameters	Values
Height (m)	12
Floor area (m^2)	390.5
Volume (m^3)	1171.5
Façade surface (m^2)	667.5
Window surface (m^2)	41

Table 2. U-values of the basic building envelope components.

Components	U-Values (W/m^2K)
External wall	0.6
Floor	0.85
Roof	0.55
Windows	2.4

Considering the beforementioned values for the building examined in this study, an energy simulation tool, called TEE KENAK, was used for calculating heating/cooling and DHW energy demand on a monthly basis, excluding energy systems. This tool follows the methodology provided by Greek EPBD, considering simulation data and parameters as set in [39]. Figure 3 presents the results of the simulation, highlighting that cooling demand is higher than the heating demand. This is due to the climatic conditions and the building thermal properties considered, as well as due to the assumption of no shading during the year. Also, defining the power demand for sizing the energy systems for each energy use is an important factor in building energy analysis. For this reason, the worst-case scenario for the climatic conditions was considered, calculating 25.1 kW for heating and 16.3 kW for cooling. Similarly, the power for the DHW is calculated at 1.8 kW, assuming 5 h of system daily operation to fully meet the needs, as defined in the technical guidelines provided in [39].

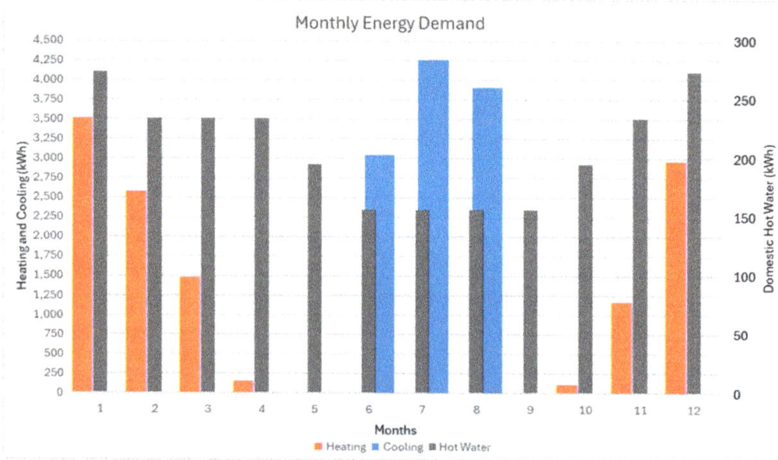

Figure 3. Building energy demand for space heating, cooling and hot water uses on a monthly basis.

2.3. Multi-Energy Systems Parameters

The proposed multi-energy systems configuration is focused on enhancing the RES penetration for meeting building energy demand. In fact, air–water heat pumps, solar thermal collectors and photovoltaic systems were included, as well as the interconnected electricity grid. It is noted that the analysis excludes the impact of different terminal units, as well as the temperature of the working medium, on energy recovery efficiency, assuming that the proposed solutions are capable of meeting the energy requirements.

For each system and their energy sources, the basic techno-economic, energy and environmental parameters, considering life cycle aspects as well, should be clarified. More specifically, the economic parameters include the operational costs provided by the consumption of the respective energy sources, as well as the costs of installing the proposed energy systems. In addition, the energy parameters account for the primary energy factors, considering the origin of the energy sources during energy system operation. Moreover, they take the embodied energy of the systems into account. This corresponds to the cumulative energy consumption originating from fossil fuels and RESs across several life cycle stages up to the system construction. Similarly, the environmental aspects incorporate the equivalent CO_2 emissions during the operation of the energy systems and the embodied emissions coming from the construction life cycle stages. The estimated impact values of the embodied energy and emissions of the examined energy systems resulted from a life cycle assessment (LCA) analysis conducted in [40,41]. In these studies, the construction of the inventory database was made by utilizing the "Ecoinvent" database and the Environmental Product Declaration (EPD) [42]. Also, the "CML 2 Baseline 2000" and cumulative energy demand (CED) methods were implemented for defining the results of energy and environmental impacts [43,44].

Regarding the technical characteristics of the energy systems examined, their lifetime duration and their efficiency were considered. For the calculation of the coefficient of performance (COP) of the heat pump, an empirical equation was used, as provided in [45]:

$$\mathrm{COP} = 0.001 \cdot \Delta \mathrm{T}^2 - 0.1534 \cdot \Delta \mathrm{T} + 7.3775, \quad (1)$$

where ΔT represents the difference between outdoor air temperature and the outlet temperature of the working medium (water), which is assumed to be 35 °C and 15 °C for heating and cooling, respectively; underfloor heating and cooling are considered for attributing thermal energy to the building.

Also, the f-chart method, developed by Duffie and Beckman, was utilized for sizing a flat plate solar thermal collector (Table 3). This approach is widely used to evaluate the annual thermal performance of active heating systems in buildings, as it offers reliable results and has a relatively easy implementation; the minimum temperature of energy delivery is around 20 °C [46]. In more detail, it provides a fully developed methodology for calculating the solar thermal fraction of meeting heating and DHW demand for a specific solar heating system. There are some key variables considered as essential for the f-chart method. The primary design variable seems to be the area of the solar collector, while the secondary variables include the collector types, storage capacity, fluid flow rates and collector heat exchanger sizes. Two dimensionless variables (X, the ratio of collector losses to heating loads and Y, the ratio of absorbed solar radiation to heating loads), resulting from several correlations, were introduced, in order for the fraction of the monthly heating load supplied by solar energy to be estimated.

Table 3. Basic design parameters of a flat plate solar thermal collector [47].

Solar Collector Parameters	Values
Thermal performance curve slope—F_RU_L	4 W/m^{2}K
Thermal performance intercept—$F_R(\tau\alpha)_n$	0.75
Coefficient for collector location—$\underline{\tau\alpha}/(\tau\alpha)_n$	0.96
Heat exchanger coefficient—F'_R/F_R	0.95
Storage tank volume per collector area—M	75 L/m^2

It should be mentioned that the f-chart method primarily applies to systems designed to meet DHW demand and a small percentage of heating demands, as defined in [46]. However, in the proposed analysis, it was assumed that the f-chart could also predict heating operation on a larger scale, considering the presence of underfloor heating, setting a balance between the temperature levels of demand (load) and supply (collector operation).

Surrogate polynomial models were developed in ALAMO (Automatic Learning of Algebraic Models) [48] for simulating the operation of the solar thermal collector, considering the results of the f-chart implementation. More specifically, these models, which are essential for the optimization process, calculate the collector area as a function of the fraction of the heating load supplied by solar energy. The process of creating such surrogate models was performed to provide models that maximize R^2, which represents the variability explained by the model, considering the total number of deviations and minimizing the root mean square error (RMSE).

Considering all of the above, the technical, economic, energy and environmental parameters utilized in the proposed study were presented in Tables 4 and 5. It is noted that a sensitivity analysis was conducted for the parameters dealing with the electricity grid, considering the share of RES technologies in the energy mixture. In more detail, the primary energy factor of 2.1 and the GHG emission factor of 0.6 kg CO_2/kWh correspond to an energy mixture of Greece in 2019, with a share of 33.2% RESs, while the values of 1.8 and 0.2 kg CO_2/kWh, respectively, refer to an increased penetration of RESs in the energy mixture, as anticipated for the future (years 2028–2030) [49–51].

Table 4. Economic, energy and environmental operational costs for the energy sources [39,49–51].

Energy Sources	Economic Costs (EUR/kWh)	Primary Energy Factors	GHG Emission Coefficients (kg CO_2/kWh)
Solar Energy	0	0	0
Electrical Energy	0.12/0.19	2.9/2.1/1.8	0.989/0.6/0.2

Table 5. Technical, economic, energy and environmental parameters for energy systems.

Energy System Parameters	Heat Pump	Solar Collector	Photovoltaic System
Efficiency	4/4.3 [1]	f-chart	0.2
Life Duration (years)	25	25	25
Installation Costs	600 EUR/kW	380 EUR/m^2	250 EUR/m^2
Energy Costs	142 kWh/kW	103 kWh/m^2	192.5 kWh/m^2
Environmental Costs	337 kg CO_2/kW	1890 kg CO_2/m^2	1022 kg CO_2/m^2

[1] The values represent the COP and EER for the heat pumps, respectively.

2.4. Energy Hub Formulation

The multi-energy systems described above were used in order to connect the energy sources with the building energy demand. This link is essential for determining the energy consumption, leading to economic, energy and environmental costs. For this reason, the EH concept was utilized, representing such connections, with its main goal being to provide optimal decisions considering the installation and operation of the energy systems under the objectives of minimizing economic, energy and environmental life cycle costs. In more detail, the formulation of the proposed EH includes the energy sources as input parameters and building energy demand as output parameters. The energy systems compose the converters in the EH concept.

This concept is illustrated in Figure 4, where electrical and solar energy are proposed to supply multi-energy systems for meeting building demand for space heating, cooling and DHW. More specifically, solar thermal collectors are proposed to cover heating or DHW demand, while heat pumps could be used for space heating, cooling and hot water. The electrical energy consumed by the heat pumps can be supplied either by photovoltaic systems or by the electricity coming from the grid. All of these decisions describe an optimization problem for defining the participation level of each of these systems in order to meet the relative building energy requirements. The optimal decisions for each criterion are made under an annual analysis, providing solutions for the systems' participation to fully cover the energy demand of the case study building.

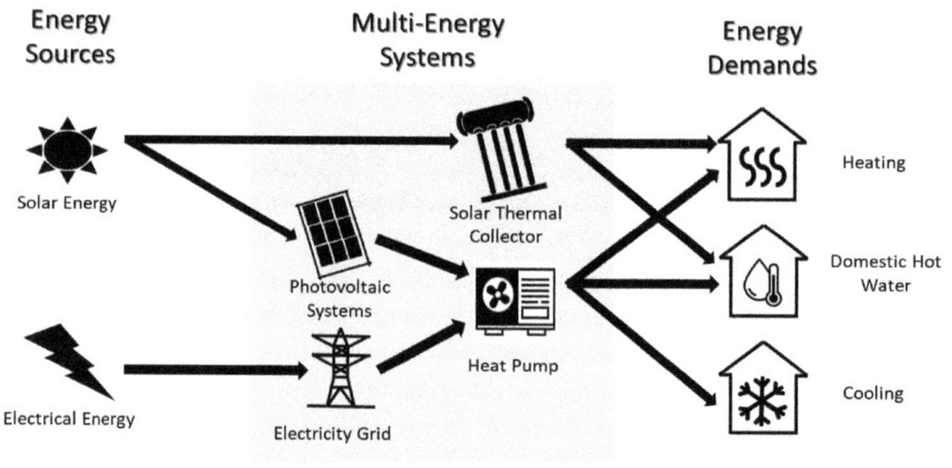

Figure 4. Description of the proposed building energy hub.

2.5. Optimization Problem

The formulation of such an EH, including multi-energy systems, creates an optimization problem for providing decisions on energy system installation and operation for each energy use. In this context, an MP problem was developed in order to find the optimal solutions by defining the relationship between key variables and constraints for achieving specific objectives, like minimizing the economic, energy and environmental costs of the systems. The following aspects are essential for MP model formulation, as also presented in Figure 5.

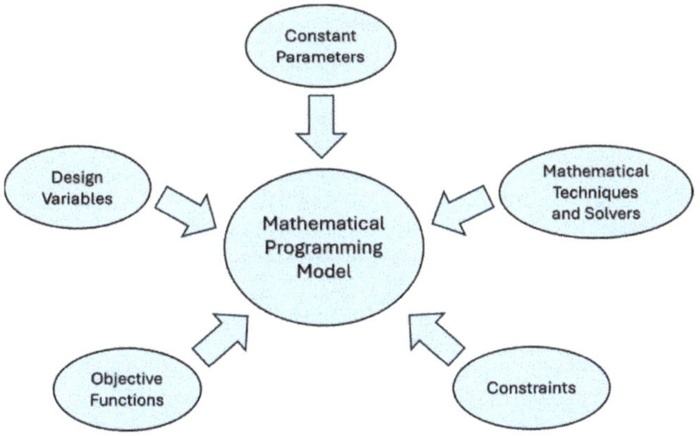

Figure 5. Description of the basic aspects that formulate an MP model.

1. Constant parameters include all of the fixed values that are considered to be unchanged when solving the optimization problem.
2. Design variables describe the optimization factors for the decision-making process and providing optimal solutions.
3. Objective functions formulated by mathematical expressions include design variables and determining the goals of the optimization problem, i.e., the optimization criteria.
4. Constraints impose the boundaries or the limitations or the requirements of the design variable in the optimization problem.
5. Mathematical techniques define the type of the optimization problem (linear, integer, etc.) and select the appropriate solver for finding the optimal solutions.

2.5.1. Constant Parameters

These parameters include building energy demand and power, as presented in Section 2.2, as well as all of the technical, economic, energy and environmental aspects for the energy sources and the systems included in Section 2.3. In more detail, economic costs, primary energy factors and GHG coefficients were depicted in Table 4 for electricity, defining the operation costs for each optimization criterion, combined with energy consumption. Also, Table 5 presents the values of installation costs, embodied energy and embodied GHG emissions of the examined energy systems. Such values correspond to the installation costs for each optimization criterion, combined with the size of the energy systems (in kW or m^2). Such values describe the optimization goals, formulating the objective functions for the economic, energy and environmental criteria.

2.5.2. Design Variables

The optimal decisions are made according to the results provided by the design variables. The key variables in this problem provide the energy systems' participation (ε_i^j), considering their proper operation and installation for meeting building energy demand. More specifically, these variables are related to the operation of the heat pumps connected to the electricity grid, as well as to the sizing of the solar thermal collectors and the photovoltaic systems. So, the range value of these variables is (0,1). However, the sizing of the heat pumps should be defined by a binary variable ($b^j_{i=HP}$) in order for the full demand size to be considered. The beforementioned design variables and the referring indexes are presented in Table 6.

Table 6. Definition of the design variables.

Design Variables	Values	Definition
ε^j_i	Non-negative [0,1]	The participation level of the energy systems' operation
$b^j_{i=HP}$	Binary (0 or 1)	The participation of the energy systems' installation (HP)
i represents the energy systems examined, i.e., heat pump (HP), solar thermal collector (SC) and the photovoltaic system (PV)		
j represents the energy uses; thus, the building energy demand for space heating (H), cooling (C) and domestic hot water (DHW)		

2.5.3. Objective Functions

The optimization criteria investigated in this study focus on minimizing the economic, energy and environmental parameters, as analysed in Section 2.3, considering heating, cooling and hot water energy demand. The form of the objective functions used in this study is presented in Equation (2).

$$Min\begin{bmatrix} Cost \\ Ener \\ Envir \end{bmatrix} = \Sigma^{j}_{i=PV}\left(\begin{bmatrix} InstCost \\ EmbEner \\ GHGEm \end{bmatrix}_i \cdot \frac{1}{LD_i} \cdot \frac{\frac{Q^j_{dem}}{COP^j}}{n_{PV}\cdot H_{sol}} \cdot \varepsilon^j_i\right) +$$
$$\Sigma^{j}_{i=HP}\left(\begin{bmatrix} OpCost \\ PrEner \\ OpGHGEm \end{bmatrix}_i \cdot \frac{Q^j_{dem}}{COP^j} \cdot \varepsilon^j_i\right) + \Sigma^{j}_{i=HP}\left(\begin{bmatrix} InstCost \\ EmbEner \\ GHGEm \end{bmatrix}_i \cdot \frac{1}{LD_i} \cdot P^j_{dem} \cdot b^j_i\right) + \quad (2)$$
$$\Sigma^{j=H,HW}_{i=SC}\left(\begin{bmatrix} InstCost \\ EmbEner \\ GHGEm \end{bmatrix}_i \cdot \frac{1}{LD_i} \cdot f\left(\varepsilon^j_i\right)\right),$$

where

- *Cost* (EUR): The total annual economic costs including systems' operation and installation, which are set for minimization.
- *Ener* (kWh): The total annual energy costs including systems' primary energy consumption and embodied energy, which are set for minimization.
- *Envir* (kg CO_2): The total annual environmental costs including systems' GHG emission during operation and the embodied emissions, which are set for minimization.
- *InstCost* (EUR/kW or m^2): The installation costs of the systems (Table 5).
- *EmbEner* (kWh/kW or m^2): The embodied energy of the systems (Table 5).
- *GHGEm* (kg CO_2/kW or m^2): The embodied GHG emissions of the systems (Table 5).
- *OpCost* (EUR/kWh): The economic costs resulting from the systems' operation (Table 4).
- *PrEner* (-): The primary energy factors for the grid electricity (Table 4).
- *OpGHGEm* (kg CO_2/kWh): The environmental costs, i.e., the GHG emitted during systems operation (Table 4).

- LD_i (years): The life duration of the energy systems examined (Table 5).
- $Q^j{}_{dem}$ (kWh): The building energy demand for each energy use (j).
- $P^j{}_{dem}$ (kW): The power demand for each energy use (j).
- H_{sol} (kWh/m^2): The incident solar irradiation.
- n_{PV}, COP^j: The efficiency of the photovoltaic systems and the coefficient of performance of the heat pumps (Table 5).

2.5.4. Constraints

For finding feasible solutions, the design variables should be constrained, considering the following aspects.

- The design variable defining the participation level of the energy systems should be limited to a lower bound of 0% and an upper bound of 100%.

$$0 \leq \varepsilon_i^j \leq 1 \ \forall i, j, \quad (3)$$

- The energy demand for each energy use should be fully met by at least one energy system.

$$\sum_i \varepsilon_i^j = 1 \ \forall j, \quad (4)$$

- The installation of the heat pump should be considered only when its participation is preferable by the optimization criteria.

$$\varepsilon_{i=HP}^j \leq b_{i=HP}^j \ \forall j, \quad (5)$$

2.5.5. Mathematical Techniques

GAMS is a computational tool specialized in formulating MP problems, while it incorporates several solvers for providing optimal solutions. The proposed MP problem was modelled in GAMS, as a Mixed Integer Non-Linear Problem, and the SCIP solver was used to find optimal solutions [52].

In addition, scripts in Python 3.10.6 were developed, conducting a brute force analysis of the proposed problem. The process starts with the calculation of all of the possible combinations of the energy systems examined, with an analysis level of 0.1%. This results in almost half a million combinations, which are used for calculating economic, energy and environmental costs, describing the optimization criteria. The above brute force process of investigating the optimal solutions is directly associated with the level of analysis, which specifies the number of the alternative scenarios examined. This is the reason for making brute force analysis intractable and time consuming. However, the process described above was implemented for the optimization problem of this analysis, in order to compare the optimal solutions provided by GAMS. With the analysis level at 0.1%, the brute force analysis leads to results similar to the ones provided by GAMS.

3. Results

3.1. Optimization Results of the Proposed EH

The optimization results show that the criteria examined are not conflicting with each other. In more detail, the minimization of economic and energy costs leads to the same optimal solution, while in the environmental criterion, a small differentiation was depicted. This leads to smaller GHG emissions, but higher economic and energy costs, as presented in Figure 6.

Building energy demand for heating, cooling and DHW are fully covered by 27 kW HP combined with 25.5 m^2 PV, under net metering. This is the optimal decision considering the economic and energy criteria. When the mitigation of GHG emissions is set as the optimization objective, the previous solution of 27 kW HP is now connected with 25 m^2 PV to fully meet the heating and cooling demand, while 1 m^2 SC is preferable for meeting the DHW demand. The participation level of the SC is 30% for the DHW demand, and the rest, 70%, is covered by the HP.

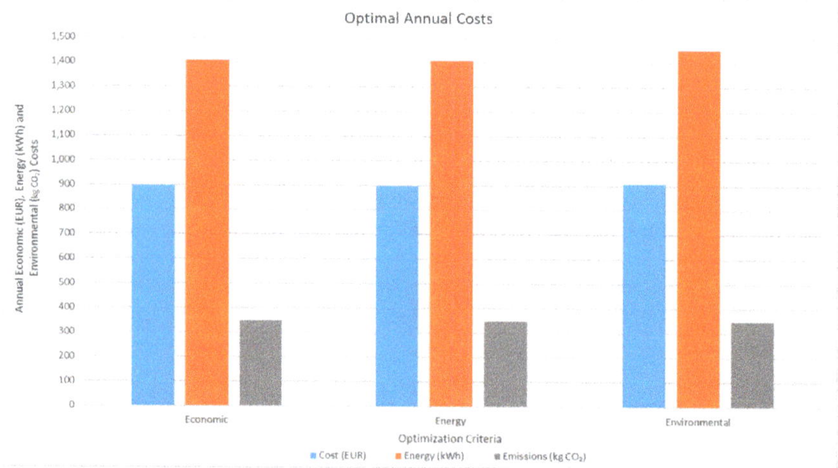

Figure 6. Optimization results for the proposed EH, considering all of the criteria examined.

Also, it is important to mention that these optimal solutions are the same when the electricity costs are reduced to 0.12 EUR/kWh, as well as when the values of primary energy and GHG coefficients are lower (2.1/0.6 or 1.8/0.2), considering a more renewable-based energy mixture for electricity [50,51].

Also, a more detailed analysis of the optimal economic values was conducted by inserting constraints in the optimization problem. Such analysis examines the basic alternative scenarios for meeting the energy demand of the building (heating/cooling/DHW), considering two values for the electricity cost. In this context, Figure 7 presents the distribution of the economic costs of these scenarios. More specifically, it appears that the minimum costs arise with the PV dominance, while the participation of SC (mainly in heating) causes an increase in costs, highlighting a great potential for economic savings. Also, economic solutions are provided when HP–PV are combined for meeting heating and cooling demand, and with SC for the DHW demand. This is due to the lower energy requirements for DHW, compared to the other two energy uses. Last but not least, the solution of HP-ElGrid seems to be economically advantageous in several cases, compared to renewable-based scenarios, especially when the price of electricity is lower.

Considering the other two criteria (energy and environmental), the distribution order for the 18 basic scenarios is different from the one in the economic criterion, especially when minimizing the GHG emissions (Figure 8). The primary energy factors and the GHG coefficient considered in this analysis are 2.1 and 0.6 kg/kWh CO_2, respectively. Results show that for the environmental criterion, the scenarios with the highest GHG emissions are those with the highest participation of the HP-ElGrid systems. While the renewable-based technologies (PV and SC) dominate, the environmental footprint decreases. The distribution order of these scenarios in the energy criterion is in line with the one in the economic criterion, apart from some exceptions of HP-ElGrid dominance that seem to be preferable.

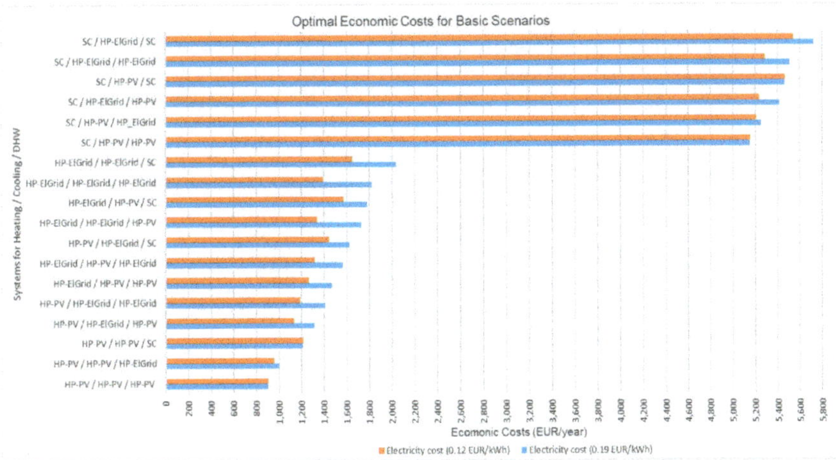

Figure 7. Optimal results of several basic scenarios, considering all the economic criterion.

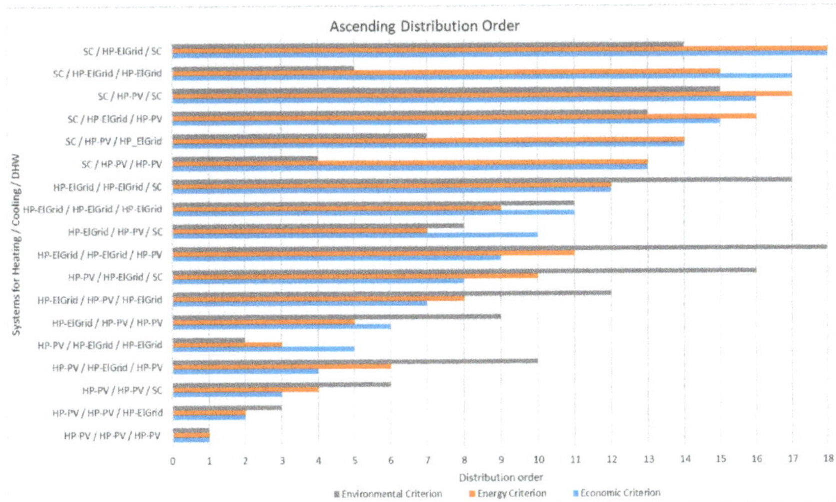

Figure 8. Ascending distribution order of the 18 basic scenarios for each optimization criterion.

3.2. Optimization Results of the EH Without PVs

The analysis was expanded to cases where PVs are totally excluded from the proposed EH, so the energy demand can be met either by SC or by HP (connected to the electricity grid) or by a combination of these two alternatives. This is important in order for a more comprehensive analysis to be carried out, providing multiple optimal results for the different criteria examined. For this reason, implementing a sensitivity analysis for the economic, energy and environmental parameters for the electricity from the grid (ElGrid), seems to be essential for the decision-making process.

In Table 7, the optimal decisions of the participation level of each energy system are presented, considering all of the criteria examined. The economic, energy and environmental parameters of the ElGrid are set as 0.19 EUR/kWh, 2.9 and 0.989 kg CO_2/kWh. Results show that the HP-ElGrid (27 kW) is dominant for the economic criterion, while when minimizing the energy and environmental costs, the participation of the SC is increased, especially for heating and DHW uses. This leads to higher economic costs, as depicted in Figure 9, which displays high conflict between the optimization criteria.

Table 7. Optimal results of energy system participation for the EH excluding PVs, considering all of the energy uses and optimization criteria.

Optimization Criteria	Economic	Energy	Environmental
Heating	100% HP-ElGrid	51.6% HP-ElGrid 48.4% SC (36 m^2)	12.6% HP-ElGrid 87.4% SC (140.5 m^2)
Cooling	100% HP-ElGrid	100% HP-ElGrid	100% HP-ElGrid
DHW	40% HP-ElGrid 60% SC (2.5 m^2)	18.8% HP-ElGrid 81.2% SC (5.5 m^2)	4.6% HP-ElGrid 95.4% SC (14.5 m^2)

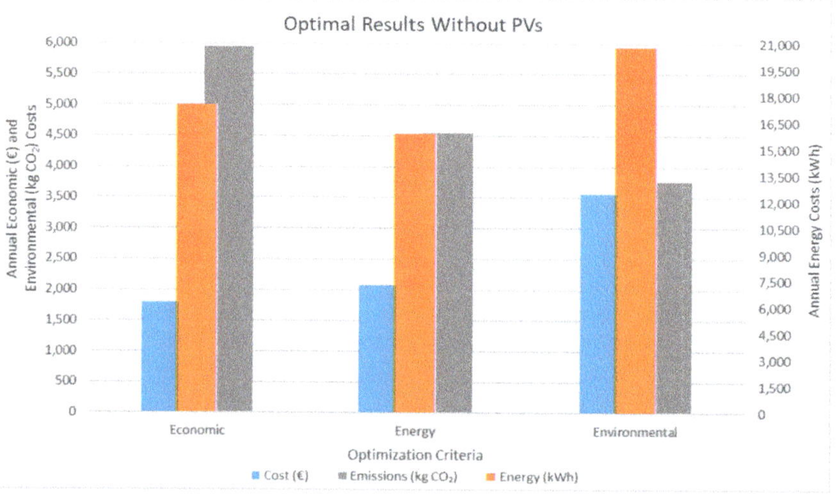

Figure 9. Optimization results for the EH excluding PVs, considering all of the criteria examined.

In addition, when decreasing the cost of electricity (0.12 EUR/kWh) in the economic criterion, the participation of SC for meeting the DHW demand drops to 41.9% (1.5 m^2). This leads to lower economic costs (23% reduction), but an increase is depicted in the energy and environmental costs (1.5–2% increase). The optimal results for the other criteria remain the same, as there is no change in energy or environmental parameters.

Considering a sensitivity analysis of the values of primary energy factors and environmental costs for the electricity, the optimization results propose minimum dependence on the SC when the energy mixture is based on RESs. This is indicative for the energy and environmental optimization criteria, leading to high participation rates of the EL-Grid. Such results are presented in Table 8, including the economic, energy and environmental costs, as well as the sizing of the energy systems installed. These are the optimal results for each optimization criterion, considering a constant value of electricity cost (0.19 EUR/kWh) and different values for primary energy factors and environmental costs. More specifically, the optimal decisions for the economic criterion are the same for the three scenarios of energy mixture; nevertheless, the energy and environmental values decrease in a renewable-based energy mixture. This pattern corresponds to the energy and environmental criteria too, but the decisions provide the mitigation of SC in renewable-based energy mixtures, which leads to lower economic costs.

Table 8. Optimal annual results of energy system participation for the EH excluding PVs, through a sensitivity analysis of primary energy factors (2.9/2.1/1.8) and environmental costs (0.989/0.6/0.2 kg CO_2/kWh), considering the (a) economic, (b) energy and (c) environmental optimization criteria.

(a) Economic Criterion				
Sensitivity Analysis Parameters	**Systems**	**Economic**	**Energy**	**Environmental**
0.19/2.9/0.989 [1]	27 kW HP-ELGrid 2.5 m² SC	EUR 1794	17,514.5 kWh	5947.6 kg CO_2
0.19/2.1/0.6 [1]			12,835.3 kWh	3672.4 kg CO_2
0.19/1.8/0.2 [1]			11,080.6 kWh	1332.7 kg CO_2
(b) Energy Criterion				
Sensitivity Analysis Parameters	**Systems**	**Economic**	**Energy**	**Environmental**
0.19/2.9/0.989 [1]	27 kW HP-ELGrid 41.5 m² SC	EUR 2082	15,870.3 kWh	4549.9 kg CO_2
0.19/2.1/0.6 [1]	27 kW HP-ELGrid 20 m² SC	EUR 1892	12,293.4 kWh	3209.8 kg CO_2
0.19/1.8/0.2 [1]	27 kW HP-ELGrid 15.5 m² SC	EUR 1855	10,776.3 kWh	1243.7 kg CO_2
(c) Environmental Criterion				
Sensitivity Analysis Parameters	**Systems**	**Economic**	**Energy**	**Environmental**
0.19/2.9/0.989 [1]	27 kW HP-ELGrid 155 m² SC	EUR 3569	20,796.4 kWh	3766.7 kg CO_2
0.19/2.1/0.6 [1]	27 kW HP-ELGrid 120.5 m² SC	EUR 3084	16,180.6 kWh	2563.1 kg CO_2
0.19/1.8/0.2 [1]	27 kW HP-ELGrid 49.5 m² SC	EUR 2172	11,448.6 kWh	1172 kg CO_2

[1] Electricity cost (EUR/kWh)/primary energy factor (-)/environmental cost (kg CO_2/kWh).

Comparing the alternative optimal solutions of Table 7 and the solution with the participation of PVs (Section 3.1), it is important to mention that when minimizing the economic costs, the optimal solution without PVs is the one provided by renewable-based energy mixtures. This leads to low values of energy and GHG emissions.

4. Conclusions

In this study, a decision-making methodology has been proposed, considering building energy design. More specifically, the EH concept was implemented in a building in order to optimize multi-energy system installation and operation in the scope of building energy demand satisfaction. The optimization was conducted under the principles of MP, considering economic, energy and environmental criteria with a life cycle perspective.

The analysis showed that the application of the optimization process in the design of multi-energy systems highlights the investigation of multiple parameters, such as technical, economic, energy and environmental ones. Also, different limitations upon the energy sources can be easily considered, providing the possibility of evaluating the optimal results too. Last but not least, the concept of EH is important for incorporating different simulation models, like f-chart, in order to provide a more accurate analysis. Additionally, the optimization with MP seems to be less time consuming than brute force investigations, where a huge number of alternative scenarios should be considered.

The proposed RES-based EH, which is considered as a tool of energy management, highlights the ability of buildings to be prosumers, thus consumers and producers concurrently. This is indicative for the optimal results, as the combination of heat pumps and photovoltaic systems under net metering seems to be preferable for all of the examined criteria. However, in cases in which the economic costs or the primary energy factors of the electricity from the grid were decreased, the optimal decisions of HP and PV remained. This is a useful conclusion for highlighting the importance of buildings' self-energy production, rather than transforming the energy mixture for electricity, which is essential in cases where the installation of PVs on buildings is forbidden. More specifically, the sensitivity analysis for the primary energy factors in an EH where the PVs were excluded showed that with more RESs in the energy mix, there were lower energy and environmental costs.

All in all, optimization models can enhance and streamline the building design process by assessing the advantages and drawbacks of different energy systems, enabling a comparative analysis of the available alternatives too. The proposed methodology can easily be implemented to different building types and climatic condition scenarios (future research), affecting the values of the economic, energy and environmental costs, upon the same pattern of optimal decisions. The current analysis was limited to an annual analysis for decision-making, considering data on a monthly basis. Future research could extend the proposed methodology to a shorter time-step analysis (daily and hourly), enabling the prediction of the dynamic operational behaviour of the energy systems, considering their efficiency degradation, especially for PV systems, and a more comprehensive strategy for their operation. In this context, it is important to use relevant approaches dedicated to space heating, especially for solar collector operation, that are based on validated simulation results. This is a meaningful task for future work, indicating a suitable form for integrating such approaches into the energy hub optimization concept. The incorporation of different energy storage technologies, such as batteries, in the energy hub, seems to be crucial too. Moreover, scenarios incorporating the fluctuation in energy demand and considering the indoor temperature setting could enhance the resilience of the proposed analysis. Lastly, thermal comfort parameters could also be examined in more detail, enhancing the importance of implementing multi-objective optimization methodologies through conflicting criteria scenarios.

Author Contributions: Conceptualization, V.K. and G.P.; methodology, V.K. and G.A.; software, N.P.; validation, V.K. and G.A.; formal analysis, V.K., N.P. and G.P.; investigation, V.K. and G.A.; resources, G.P.; data curation, V.K. and G.A.; writing—original draft preparation, V.K.; writing—review and editing, N.P. and G.P.; visualization, V.K.; supervision, N.P. and G.P.; project administration, G.P.; funding acquisition, V.K., N.P. and G.P. All authors have read and agreed to the published version of the manuscript.

Funding: This research was funded by the Hellenic Foundation for Research and Innovation (HFRI), under the 4th call for HFRI PhD Fellowships (Fellowship Number 11142).

Data Availability Statement: The original contributions presented in this study are included in the article. Further inquiries can be directed to the corresponding author.

Conflicts of Interest: The authors declare no conflicts of interest. The funders had no role in the design of the study; in the collection, analyses, or interpretation of data; in the writing of the manuscript; or in the decision to publish the results.

Abbreviations

The following abbreviations are used in this manuscript:

EU	European Union
GHG	greenhouse gas
RES	renewable energy source
EED	Energy Efficiency Directive
EPBD	Energy Performance of Buildings Directive
RED	Renewable Energy Directive

GOV	Governance Regulation
NECPs	National Energy and Climate Plans
NZEB	nearly zero-energy building
ZEB	zero-emission building
MEPS	Minimum Energy Performance Standards
ESM	energy saving measures
ESS	energy supply systems
EH	energy hub
LCA	life cycle assessment
GAMS	General Algebraic Modelling System
MP	mathematical programming
U-values	thermal transmittance coefficients
DHW	domestic hot water
H	space heating
C	space cooling
CO_2	carbon dioxide
EPD	Environmental Product Declaration
CED	cumulative energy demand
COP	coefficient of performance
EER	energy efficiency ratio
ALAMO	Automatic Learning of Algebraic Models
RMSE	root mean square error
HP	heat pump
SC	solar thermal collector
PV	photovoltaic system
LD	life duration

References

1. International Energy Agency (IEA). *World Energy Outlook. Secure Sustainable Together*; International Energy Agency: Paris, France, 2018.
2. Eurostat. Interactive Publications. Shedding Light on Energy in Europe—2024 Edition. Available online: https://ec.europa.eu/eurostat/web/interactive-publications/energy-2024#energy-consumption (accessed on 19 March 2025).
3. Hamilton, I.; Kennard, H. *2020 Global Status Report for Buildings and Construction: Towards a Zero-Emission, Efficient and Resilient Buildings and Construction Sector*; United Nations Environment Programme: Nairobi, Kenya, 2020.
4. Wong, J.K.W.; Li, H. Intelligent building research: A review. *Autom. Constr.* **2005**, *14*, 143–159. [CrossRef]
5. European Commission, Directorate-General for Energy. EU Energy in Figures Statistical Pocketbook 2023. Publications Office of the European Union. 2023. Available online: https://data.europa.eu/doi/10.2833/502436 (accessed on 19 March 2025).
6. Eurostat. Energy Statistics—An Overview. 2021. Available online: https://ec.europa.eu/eurostat/statistics-explained/index.php?title=Energy_statistics_-_an_overview#Primary_energy_production (accessed on 19 March 2025).
7. UN Environment and IEA. Towards a Zero-Emission, Efficient, and Resilient Buildings and Construction Sector. Global Status Report 2017. 2017. Available online: www.globalabc.org (accessed on 1 March 2019).
8. Rosenow, J.; Cowart, R.; Bayer, E.; Fabbri, M. Assessing the European Union's Energy Efficiency Policy: Will the Winter Package deliver on "Efficiency First"? *Energy Res. Soc. Sci.* **2017**, *26*, 72–79. [CrossRef]
9. Buildings Performance Institute Europe. 2017. Available online: https://www.bpie.eu/publication/97-of-buildings-in-the-eu-need-to-be-upgraded/ (accessed on 19 March 2025).
10. European Environmental Agency (EEA). Greenhouse Gas Emissions from Energy Use in Buildings in Europe. 2024. Available online: https://www.eea.europa.eu/en/analysis/indicators/greenhouse-gas-emissions-from-energy (accessed on 19 March 2025).
11. Buildings Performance Institute (BPIE). Deep Renovation: Shifting from Exception to Standard Practice in EU Policy. 2021. Available online: https://www.bpie.eu/wp-content/uploads/2021/11/BPIE_Deep-Renovation-Briefing_Final.pdf (accessed on 19 March 2025).
12. International Renewable Energy Agency (IRENA). World Energy Transitions Out-Look. 2022. Available online: https://www.irena.org/Digital-Report/World-Energy-Transitions-Outlook-2022 (accessed on 19 March 2025).
13. Buildings Performance Institute (BPIE). Report on the Evolution of the European Regulatory Framework for Buildings Efficiency. 2022. Available online: https://www.bpie.eu/wp-content/uploads/2022/02/rev6_SPIPA_EU.pdf (accessed on 19 March 2025).

14. European Union. Directive (EU) 2024/1275 of the European Parliament and of the Council of 24 April 2024 on the Energy Performance of Buildings (Recast). In *Official Journal of the European Union*; European Union: Brussels, Belgium, 2024.
15. European Commission. Communication from the Commission to the European parliament, the European Council, the Council, the European Economic and Social Committee and the Committee of the Regions the European Green Deal. Brussels. 2019. Available online: https://eur-lex.europa.eu/legal-content/EN/TXT/?uri=COM:2019:640:FIN (accessed on 19 March 2025).
16. European Commission. Commission Welcomes Completion of Key "Fit for 55" Legislation, Putting EU on Track to Exceed 2030 Targets. Brussels. 2023. Available online: https://ec.europa.eu/commission/presscorner/detail/en/ip_23_4754 (accessed on 19 March 2025).
17. European Commission. Communication from the Commission to the European Parliament, the European Council, the Council, the European Economic and Social Committee and the Committee of the Regions. REPowerEU Plan. Brussels. 2022. Available online: https://eur-lex.europa.eu/legal-content/EN/TXT/?uri=COM:2022:230:FIN&qid=1653033742483 (accessed on 19 March 2025).
18. Zhang, C.; Cui, C.; Zhang, Y.; Yuan, J.; Luo, Y.; Gang, W. A review of renewable energy assessment methods in green building and green neighborhood rating systems. *Energy Build.* **2019**, *195*, 68–81. [CrossRef]
19. Ran, S.; Lyu, W.; Li, X.; Xu, W.; Wang, B. A solar-air source heat pump with thermosiphon to efficiently utilize solar energy. *J. Build. Eng.* **2020**, *31*, 101330. [CrossRef]
20. Nik-Bakht, M.; Panizza, R.O.; Hudon, P.; Chassain, P.Y.; Bashari, M. Economy-energy trade off automation—A decision support system for building design development. *J. Build Eng.* **2020**, *30*, 2352–7102. [CrossRef]
21. Fabrizio, E.; Corrado, V.; Filippi, M. A model to design and optimize multi-energy systems in buildings at the design concept stage. *Renew. Energy* **2010**, *35*, 644–655. [CrossRef]
22. Kilis, V.; Paschalidis, N.; Ploskas, N.; Panaras, G. Energy Hub Optimization on Residential Building Case. *E3S Web Conf.* **2023**, *436*, 01017. [CrossRef]
23. Li, A.; Xiao, F.; Zhang, C.; Fan, C. Attention-based interpretable neural network for building cooling load prediction. *Appl. Energy* **2021**, *299*, 117238. [CrossRef]
24. Haddad, S.; Synnefa, A.; Marcos, M.Á.P.; Paolini, R.; Delrue, S.; Prasad, D.; Santamouris, M. On the potential of demand-controlled ventilation system to enhance indoor air quality and thermal condition in Australian school classrooms. *Energy Build.* **2021**, *238*, 110838. [CrossRef]
25. Zhang, C.; Gang, W.; Wang, J.; Xu, X.; Du, Q. Numerical and experimental study on the thermal performance improvement of a triple glazed window by utilizing low-grade exhaust air. *Energy* **2019**, *167*, 1132–1143. [CrossRef]
26. D'Agostino, D.; Mazzarella, L. What is a Nearly zero energy building? Overview, implementation and comparison of definitions. *J. Build. Eng.* **2019**, *21*, 200–212. [CrossRef]
27. Harkouss, F.; Fardoun, F.; Biwole, P.H. Multi-objective optimization methodology for net zero energy buildings. *J. Build. Eng.* **2018**, *16*, 57–71. [CrossRef]
28. Longo, S.; Montana, F.; Riva Sanseverino, E. A review on optimization and cost-optimal methodologies in low-energy buildings design and environmental considerations. *Sustain. Cities Soc.* **2019**, *45*, 87–104. [CrossRef]
29. Ferrara, M.; Della Santa, F.; Bilardo, M.; De Gregorio, A.; Mastropietro, A.; Fugacci, U.; Vaccarino, F.; Fabrizio, E. Design optimization of renewable energy systems for NZEBs based on deep residual learning. *Renew. Energy* **2021**, *176*, 590–605. [CrossRef]
30. Zhang, S.; Pan, G.; Li, B.; Gu, W.; Fu, J.; Sun, Y. Multi-Timescale Security Evaluation and Regulation of Integrated Electricity and Heating System. *IEEE Trans. Smart Grid* **2024**, *16*, 1088–1099. [CrossRef]
31. Sergi, B.; Pambour, K. An Evaluation of Co-Simulation for Modeling Coupled Natural Gas and Electricity Networks. *Energies* **2022**, *15*, 5277. [CrossRef]
32. Mancarella, P. MES (multi-energy systems): An overview of concepts and evaluation models. *Energy* **2014**, *65*, 1–17. [CrossRef]
33. Sontag, R.; Lange, A. Cost effectiveness of decentralized energy supply systems taking solar and wind utilization plants into account. *Renew. Energy* **2003**, *28*, 1865–1880. [CrossRef]
34. Zhao, R.; Zheng, W. Efficient operation of combined residential and commercial energy hubs incorporating load management and two-point approximation for uncertainty modeling. *Comput. Electr. Eng.* **2024**, *116*, 109197. [CrossRef]
35. Zhong, S.; Wang, X.; Wu, H.; He, Y.; Xu, B.; Ding, M. Energy hub management for integrated energy systems: A multi-objective optimization control strategy based on distributed output and energy conversion characteristics. *Energy* **2024**, *306*, 132526. [CrossRef]
36. Zhai, X.; Li, Z.; Li, Z.; Xue, Y.; Chang, X.; Su, J.; Jin, X.; Wang, P.; Sun, H. Risk-averse energy management for integrated electricity and heat systems considering building heating vertical imbalance: An asynchronous decentralized approach. *Appl. Energy* **2025**, *383*, 125271. [CrossRef]
37. Tushar, M.H.K.; Zeineddine, A.W.; Assi, C. Demand-side management by regulating charging and discharging of the EV ESS and utilizing renewable energy. *IEEE Trans. Ind. Inform.* **2018**, *14*, 117–126. [CrossRef]

38. Li, Z.; Wu, L.; Xu, Y.; Zheng, X. Stochastic-Weighted Robust Optimization Based Bilayer Operation of a Multi-Energy Building Microgrid Considering Practical Thermal Loads and Battery Degradation. *IEEE Trans. Sustain. Energy* **2022**, *13*, 668–682. [CrossRef]
39. *TEE 20701-1*; Analytical Technical Specifications of Parameters for the Calculation of Buildings' Energy Performance and the Issuing of Energy Performance Certificate. Technical Directive of the Technical Chamber of Greece (TEE): Athens, Greece, 2017.
40. Manoudis, A. Life Cycle Analysis of Energy Systems Used in Residential Buildings. Master's Thesis, International Hellenic University, Thessaloniki, Greece, 2011.
41. Longo, S.; Cellura, M.; Guarino, F.; La Rocca, V.; Maniscalco, G.; Morale, M. Embodied energy and environmental impacts of a biomass boiler: A life cycle approach. *AIMS Energy* **2015**, *3*, 214–226. [CrossRef]
42. Frischknecht, R.; Rebitzer, G. The Ecoinvent database system: A comprehensive web based LCA database. *J. Clean. Prod.* **2005**, *13*, 1337–1343. [CrossRef]
43. Cabal, H.; Lechon, Y.; Saez, R. Environmental aspects of integration of decentralized generation into the overall electricity generation system. In *EUSUSTEL Task Report*; Ministerio de Educación y Ciencia: Madrid, Spain, 2005.
44. Ochoa, L.; Ries, R.; Matthews, H.S.; Hendrickson, C. Life Cycle Assessment of Residential Buildings. In *American Society of Civil Engineers Construction Research Congress*; Working Paper; American Society of Civil Engineers: Reston, VA, USA, 2005.
45. Mouzeviris, G.; Papakostas, K. Comparative Analysis of Air-to-Water and Ground Source Heat Pumps Performances. *Int. J. Sustain. Energy* **2021**, *40*, 69–84. [CrossRef]
46. Duffie, J.A.; Beckman, W.A.; Blair, N. *Solar Engineering of Thermal Processes, Photovoltaics and Wind*, 5th ed.; Wiley: Hoboken, NJ, USA, 2020.
47. Rosli, M.A.M.; Zaki, D.S.M.; Rahman, F.A.; Sepeai, S.; Nurfaizey, A.H.; Nawam, M. F-chart method for design domestic hot water heating system in Ayer Keroh Melaka. *J. Adv. Res. Fluid Mech. Therm. Sci.* **2019**, *56*, 59–67.
48. Wilson, Z.T.; Sahinidis, N.V. The ALAMO approach to machine learning. *Comput. Chem. Eng.* **2017**, *106*, 785–795. [CrossRef]
49. Bilard, M.; Galata, S.; Fabrizio, E. The role of Primary Energy Factors (PEF) for electricity in the evaluation and comparison of building energy performance: An investigation on European nZEBs according to EN 17423:2020. *Sustain. Cities Soc.* **2019**, *87*, 104189. [CrossRef]
50. Balaras, C.A.; Dascalaki, E.G.; Psarra, I.; Cholewa, T. Primary Energy Factors for Electricity Production in Europe. *Energies* **2023**, *16*, 93. [CrossRef]
51. Balaras, C.A.; Dascalaki, E.G.; Patsioti, M.; Droutsa, K.G.; Kontoyiannidis, S.; Cholewa, T. Carbon and Greenhouse Gas Emissions from Electricity Consumption in European Buildings. *Buildings* **2023**, *14*, 71. [CrossRef]
52. Achterberg, T. SCIP: Solving constraint integer programs. *Math. Prog. Comp.* **2009**, *1*, 1–41. [CrossRef]

energies

MDPI

Article

Optimization of Hydronic Heating System in a Commercial Building: Application of Predictive Control with Limited Data

Rana Loubani [1,*], Didier Defer [1], Ola Alhaj-Hasan [1] and Julien Chamoin [2]

[1] Univ. Artois, ULR 4515, Laboratoire de Génie Civil et géo-Environnement (LGCgE), F-62400 Béthune, France; didier.defer@univ-artois.fr (D.D.); ola.alhajhasan@univ-artois.fr (O.A.-H.)

[2] Junia, ULR 4515, Laboratoire de Génie Civil et géo-Environnement (LGCgE), F-59000 Lille, France; julien.chamoin@junia.com

* Correspondence: rana.loubani@univ-artois.fr

Abstract: Optimizing building equipment control is crucial for enhancing energy efficiency. This article presents a predictive control applied to a commercial building heated by a hydronic system, comparing its performance to a traditional heating curve-based strategy. The approach is developed and validated using TRNSYS18 modeling, which allows for comparison of the control methods under the same weather boundary conditions. The proposed strategy balances energy consumption and indoor thermal comfort. It aims to optimize the control of the secondary heating circuit's water setpoint temperature, so it is not the boiler supply water temperature that is optimized, but rather the temperature of the water that feeds the radiators. Limited data poses challenges for capturing system dynamics, addressed through a black-box approach combining two machine learning models: an artificial neural network predicts indoor temperature, while a support vector machine estimates gas consumption. Incorporating weather forecasts, occupancy scenarios, and comfort requirements, a genetic algorithm identifies optimal hourly setpoints. This work demonstrates the possibility of creating sufficiently accurate models for this type of application using limited data. It offers a simplified and efficient optimization approach to heat control in such buildings. The case study results show energy savings up to 30% compared to a traditional control method.

Keywords: model predictive control; heating system optimization; artificial neural network; support vector machines; genetic algorithm

1. Introduction

1.1. Motivation

The rapid growth in energy consumption around the world has already raised supply concerns. While prices are increasing, high CO_2 emissions are causing severe climate change and extreme weather disasters. A total of 36% of the global energy is consumed by the buildings and construction sector, and it accounted for about 39% of process-related carbon dioxide emissions in 2018 [1].

To improve energy efficiency in buildings, three ways are conceivable: the first refers to the construction of high-energy-performance buildings with a low environmental footprint. But the scope of this solution is limited by a low renewal rate. According to [2], the building renewal rate is only 1.2% per year. The second is the retrofit strategy, which may lead to high material, labor, and time costs [3]. The third way to improve energy efficiency is optimizing the use of existing systems. Unlike new construction or renovations, which require significant investments, the optimization of existing systems can be achieved at a lower cost. This approach focuses on adjustment and fine-tuning, often leveraging digital solutions and control systems, eliminating the need for complete equipment replacement. Many systems, such as heating, ventilation, air-conditioning, and lighting, can be optimized to operate more efficiently. Optimizing a building's systems is a strategy that can be prioritized for a fast return on investment and tangible benefits for owners and occupants. In the field of building systems optimization, some studies have proposed the optimization of ventilation [4–6], while others have worked on the optimization of cooling systems [7–9]. In EU, HVAC systems account for 68% of the total energy consumption [10], and one of the biggest energy-consuming elements in buildings are heating systems, especially in cold and temperate countries. In France, heating systems represent 66% of the energy consumed in residential buildings and 43% in commercial buildings. In basic operation modes, as in ON/OFF mode, the building's inertia is disregarded. More elaborate methods such as PID controllers account for the building's inertia during control operations but are unable to anticipate disruptive factors like weather conditions or occupancy patterns. Both methods can thus lead to situations of discomfort and/or an excess of energy consumption. A more advanced and promising solution mentioned in the literature is the predictive control strategy or Model Predictive Control (MPC) [11].

MPC refers to a class of advanced control algorithms which predict the future response of a system based on an explicit model and optimize its future behavior [12]. The advantage of MPC lies in the ability to integrate a variety of constraints directly into its optimization process. This makes it particularly well-suited for managing complex systems like building heating. One of the MPC strategies that is often prioritized is minimizing overall energy consumption. MPC can also be configured to account for energy costs and variable pricing depending on the time of day [13,14], as well as energy curtailment constraints by reducing consumption during peak times [15] or prioritizing the use of renewable energy sources [16].

1.2. *Literature Review of MPC for Buildings' Heating Systems*

Several research studies aim to optimize the control of electric heating systems, such as electric radiators or electric heat pumps, in order to improve thermal comfort and reduce energy consumption. The control of these systems is achieved using a binary command: either ON or OFF. In [17], a predictive strategy was employed to control an electric radiator in a building under construction in France. Results showed thermal comfort in up to 67% of the hours of occupation while consuming the minimum possible energy. About 50% of energy savings were achieved in [18], where an MPC was implemented to control an electric HVAC system, covering heating during winter and cooling during summer. In [16], an MPC was implemented to reduce the electricity use by the heat pump of a heating system, favoring the use of the photovoltaic system. The study focused on optimizing energy usage remotely via cloud infrastructure. In [19], an MPC was implemented to optimize the setpoint scheduling of an electrical heating system in a building.

In this paper, the approach is different, since the proposed strategy aims to control a hydronic heating system in an office building, which represents the most common systems in Western Europe according to [20], especially systems with gas boilers combined with radiators. These boilers are the starting point of a heating circuit where the temperature of the water supplying the circuit is controlled by control valves, usually three-way valves. The setpoint temperature is transmitted to the control valves using a heating curve that depends on the outdoor temperature. In fact, the temperature of the water that is actually running in the system (the supply temperature) can deviate from the setpoint temperature given to the control valves, and that can vary from one system to another, according to the inertia and the thermal characteristics of the system. But, instead of measuring the supply water temperature, the strategy in this paper relies on the optimization of the setpoint temperature that controls the heating system.

Many research studies have been conducted for the purpose of optimizing these kinds of heating systems. Some of these works are performed with the implementation of white-box models, using EnergyPlus coupled with an Ant Colony Optimization algorithm to achieve a trade-off between computational time and solution quality [15]. Others, use grey-box models for buildings. For instance, a Model Predictive Control of heating demand in buildings connected to a district heating system is presented in [21]. Also, Chen and You [22] implemented a nonlinear MPC to control a building's climate involving renewable energy sources. These kinds of models often require detailed data about the building, which necessitates a large number of measurements. For example, an MPC using a Resistance-Capacitance thermal model was implemented in [23] for the control of a heating system in an office building which uses both floor and ceiling heating. The following measured variables were used for the control methodology: indoor temperature of each zone, supply and return temperatures, and some concrete core temperatures (heating circuits, one sensor for each). Fielsch et al. [24] implemented an MPC for hydronic heating systems in residential buildings. They also had to measure the supply and return water temperature, the temperature of the radiators, and the water mass flow rate. In [25,26], an MPC was presented for ceiling radiant heating. Both supply and return temperature had to be measured in the works, in order to build a thermal RC model; they also used the difference in supply and return temperature of the network to calculate the energy consumption of the heating system, as well as the heating degree days index (HDD), which is a quantitative index designed to reflect the demand for energy needed to heat a building. HDD was used in [27] to calculate the heating demand as well. The results of the study showed that their MPC strategy could save up to 19.69% of energy, compared to a traditional heating strategy.

Unlike the cited studies that rely on extensive measurements, this paper focuses on optimizing a heating sequence by estimating the energy demand of the proposed strategy, and this is achieved using black-box models to predict boiler gas consumption, addressing the challenge of operating with limited data. This approach was also evoked by [28], where a simplified prediction model for dynamic HVAC system loads was developed using the Modelica-based Building Library and a linear aggregation method. The objective is to minimize input requirements to basic, easily measurable parameters, addressing challenges in obtaining detailed data during regional planning.

The choice of using black-box models is based on their ability to precisely model the building, without any details or expertise in the building field. Black-box models based on machine learning are used for building modeling, such as artificial neural networks [19,27,29], support vector machines [30], random forests, or ensemble methods [31]. A predictive strategy was implemented in [29] in order to control the water flow rate of a hydronic heating system, using a Long Short-Term Memory network, a particular type of neural network. The strategy proposed outperformed the conventional PID controller in the studied building in terms of comfort, but the MPC still presented a delay in reaching the setpoint temperature at the right time. In [19], an MPC was developed using an artificial neural network for the control of scheduling heating set points in a building. The simulations showed the MPC controller achieved 27% energy reduction as compared to conventional control. Smarra et al. [32] applied random forests on real data for the prediction of indoor air temperature, but they worked on an electrical heating system. The MPC achieved up to 49.2% of energy savings compared to conventional heating strategies. An SVM model-based MPC for energy demand prediction in [33] was implemented to optimize the control variables of an urban system, to finally achieve around 37.57% of energy savings in the winter.

1.3. Objectives and Contribution of the Paper

This paper proposes a new control strategy for hydronic heating systems, focusing on optimizing the setpoint temperature of the system's control valve. This approach is innovative not only because it relies on a minimal amount of sensor data, using only information commonly available in buildings, but also because it combines two predictive models (artificial neural network (ANN) and support vector machine (SVM)). So, on one hand, there is no need for complex and expensive measurements of supply and return temperatures or water flow rates, as the strategy relies solely on the heating system setpoint temperature, which is the variable most commonly used for control purposes.

On the other hand, given the complexity of evaluating energy savings from different control strategies (as noted by Široký et al. [25]), the proposed strategy employs two predictive models. The first is an ANN that predicts the indoor air temperature, accounting for weather forecasts and occupancy. The second model is an SVM that estimates the energy consumption of the gas boiler. Both models are incorporated into the objective function of the optimization algorithm, which not only satisfies comfort constraints but also reduces energy consumption at the same time. The structure of this paper is as follows:

- Section 2 outlines the methodology of the Model Predictive Control (MPC), and its application to a case study building;
- Section 3 presents the results, and a discussion is approached in Section 4.

2. Methodology of the MPC

This paper presents a predictive strategy MPC that anticipates any disruptive event in the building, such as changes in the weather or occupancy scenarios, and optimizes the control of the heating system. Optimization is a multicriteria aspect, and therefore, it involves a compromise between perceived thermal comfort and energy consumption. To achieve this, a building model capable of predicting the building's response to different weather conditions and heating scenarios is required.

The proposed strategy is based on a black-box structure using two models of the building, where the first one predicts the indoor air temperature, and the second predicts the energy consumption. An hourly timestep is considered, which is suitable for slow-moving processes in HVAC systems [34]. As shown in Figure 1, the strategy is designed as three nested loops.

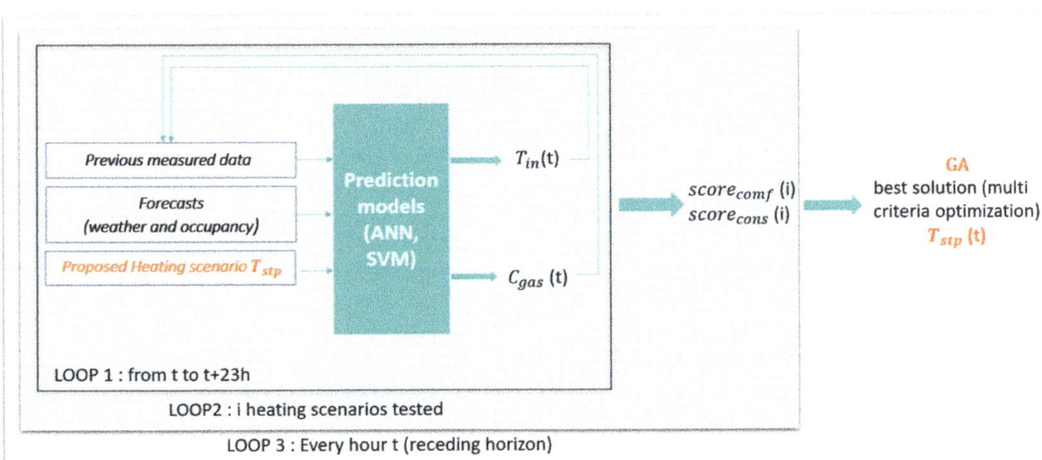

Figure 1. Predictive control strategy diagram.

1. Loop 1: Indoor air temperature and gas consumption prediction:

This loop turns for 24 iterations. It focuses on hourly predictions of the building's indoor air temperature and gas consumption over a 24-h prediction horizon (from timestep t to t + 23). In this paper, the prediction is based on machine learning models, using historical data of the building (such as indoor air temperature, outdoor air temperature, gas consumption, heating setpoint temperature, horizontal radiation, and occupancy), the weather and occupancy forecasts for the next 24 h, and the proposed heating setpoint (sequence of 24 setpoint temperature values to be optimized). Using these input data, the predicted indoor air temperature is generated by the ANN model, and the predicted boiler's gas consumption is generated by the SVM model, both over a 24-h prediction horizon. Both models are explained in Section 4.

2. Loop 2: Multicriteria optimization:

Following a defined number of iterations, and according to defined constraints, a genetic algorithm (GA) tests different heating setpoint temperature sequences and calculates a score for each proposed 24-h heating scenario. Optimizing sequences of 24 continuous values creates a large and complex search space, making genetic algorithms a justified choice for this task.

The values of the predicted indoor temperature and gas consumption vary with each tested scenario. Taking into account weather forecasts and occupancy, the genetic algorithm uses the prediction models to assess the building's response to each tested scenario (noted «i») in terms of indoor temperature and energy consumption, and consequently to calculate the final score of the objective function. The output of this second loop is the best solution found by the genetic algorithm, corresponding to the heating scenario with the best score. The calculation of this score is presented in Section 2.3.

3. Loop 3: Recalculation and receding horizon:

Once the best solution among those tested by the genetic algorithm is found, the setpoint temperature value of the first timestep T_{stp}(t) is sent to the control, the resulting indoor temperature and energy consumption values are collected to be injected back into the prediction loop (T_{in}(t), C_{gas}(t)), and the optimization is run again for a prediction horizon of 24 h, shifting by one timestep each time, forming a receding horizon of one hour. Recalculating every hour allows the predicted values to be replaced by the actual measured values of indoor temperature and gas consumption, therefore updating these values in the loop. Consequently, this avoids the accumulation of potential prediction errors throughout the prediction horizon.

Thus, the methodology can therefore be summarized as follows: the indoor air temperature and the gas consumption are predicted over a 24 h period, and every hour, this horizon moves by one timestep and the optimization is restarted in order to propose a sequence of 24 control signals for the next 24 h at each timestep.

2.1. Case Study

In order to implement the strategy and test the relevance of the method, a digital medium was used as a case study. Thus, the method presented in Section 2 was applied to data generated from a building model created in TRNSYS18 (version 18.0). A simple case of a parallelepiped building was modeled. For the architectural design, nature of the materials, heating, and control system, the model was inspired by a real building located in Lille, northern France, facing northwest. Thus, the model is conceptually based on that building, but it is not calibrated to it, and that is due to the lack of data measured on the building. In fact, the main objective remains to find a strategy that optimizes the control in the modeled building, and thus, from the perspective of applying the MPC on another building, new prediction models and controls will be elaborated.

Figure 2 shows the picture of the building that was modeled. It is a commercial building composed of three floors. Referring to standard values for buildings constructed before 2012 [21], the external walls are modeled in brick, with 6 cm of insulation, the floor in insulated reinforced concrete, and the attic insulated with 15 cm of mineral wool. Double-glazed windows are added on both sides of the building. Concerning the heating, the building is heated by a hot water radiators system, which is a very common installation in France. This type of system is often composed of two circuits: a primary and a secondary circuit [35]. Figure 3 shows the modeled heating network. The hot water heating system is based on two interconnected circuits: a primary circuit and a secondary circuit. The primary circuit mainly consists of a gas boiler, whose function is to keep the water in a buffer tank at a high temperature. The secondary circuit, on the other hand, supplies a network of radiators distributed throughout the building. To heat the building, a pump ensures the circulation of hot water in this network. The regulation of the supply water temperature, denoted as T_{stp} in Figure 3, within the secondary circuit is carried out by a three-way valve, which adjusts the mix between the return water from the secondary circuit and the hotter water coming from the buffer tank.

Figure 2. The building chosen for the case study.

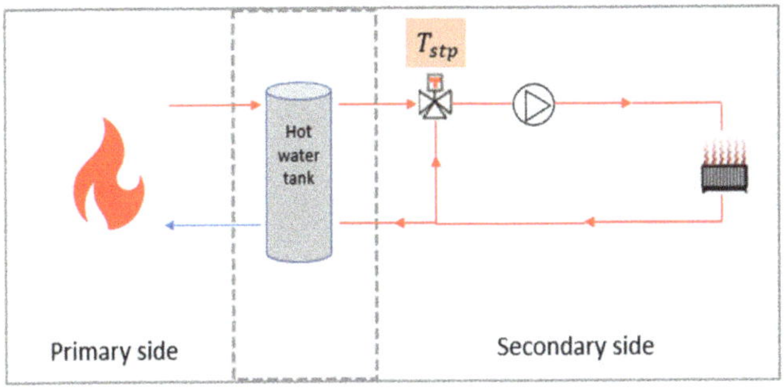

Figure 3. The modeled heating system with a primary side heat source, a hot water storage tank, and a secondary side with temperature control, circulation, and heat distribution.

In this paper, a type 526 radiator from the TESS library was used. The radiator surface was adjusted to emit the amount of heat required to heat the building. A constant flow rate was maintained on both sides of the heating network. The system adjusts the opening of the three-way valve based on the flow temperature according to the given setpoint. Thus, the setpoint temperature applied to the control valve governs the entire heating system. Buildings with this type of installation are typically controlled using a single heating curve that sets the supply temperature according to the outdoor temperature. This means that a single setpoint temperature is provided to the valves to regulate the various zones of the building. For this reason, the case study building was modeled as a single zone with a volume of 4968 m^3, controlled by a single setpoint temperature. As the authors did not have access to real data for the building, the exact values of the parameters for the heating curve of the reference building were unknown. However, theoretical considerations suggest that the heating curve is adapted to the occupancy scenario, with reduced heating at night and during unoccupied periods.

It is evident that no simple relation exists between the heating power dissipated by the radiators and the setpoint temperature controlling the 3W valve, nor between the setpoint temperature and the energy consumed by the boiler.

The TRNSYS18 simulation was performed at hourly intervals over a full year. Weather data corresponding to northern France, available in the TRNSYS18 software files, were used for the simulation. This file contains weather prediction data over a complete year. The heat provided by the radiators was obviously taken into account, as well as internal gains made by the occupants, if present. An occupancy scenario was defined from 8 a.m. to 6 p.m., Monday to Friday. This is the occupancy level corresponding to our reference building and it specifies the periods of building occupancy and non-occupancy. A ventilation scenario of 0.2 vol/h was also imposed, as well as a low infiltration.

The goal of this work was to determine an optimal hourly sequence of setpoint temperature values for the heating system's three-way control valve, over a receding prediction horizon. To simulate the building's response to varying setpoints, changing weather conditions, and occupancy patterns, a random sequence of setpoint temperatures was applied to the control valve in our building model. These values, ranging from 16 °C to 80 °C, were selected based on domain knowledge to ensure their practical relevance and were generated using a uniform random distribution. The objective was to simulate the building's response to setpoint temperatures that are independent of outdoor conditions. This aligned with the study's goal of transitioning from conventional heating curves to optimal control sequences for setpoint management. The simulation was run using these random values, generating 8760 records (one record for each hour of the year). These data were then used to model the thermal behavior of the building and heating system.

2.2. Data-Driven Prediction Models

The data used for modeling are simulation data generated by TRNSYS18, so the prediction models detailed in this section were trained using these data. The study deliberately focused on commonly measured variables, despite the availability of many others through TRNSYS18 simulations. Above all, the aim was not to measure and use the supply and the return water temperatures, but instead to model the building's response to the setpoint temperature imposed on the valve. This is the variable typically used to control this type of system. Thus, five variables represent the initial inputs of the models: setpoint temperature, outdoor temperature, horizontal radiation, indoor temperature, occupancy and gas consumption, noted, respectively, as follows: T_{cons}, T_{out}, Rad, T_{in}, Occ, and C_{gas}.

Several research studies have been carried out to model the temperature of a building zone and have shown that the most accurate prediction is obtained by adding historical data measured in the building, i.e., by integrating lagged terms of the variables as input parameters to the model. The number of lags to be used varies from one building to another and has to be studied. It depends on the inertia of each building and on the type of its systems. Fu et al. [36] worked on the prediction of a building's air-conditioning load and found the best result by taking the values of the previous 48 h of electricity consumption. Mechaqrane et al. [37] also implemented prediction models for building indoor temperature, comparing a linear autoregressive model of indoor temperature with a neural network that takes the building's past measured data as parameters. They found that a neural network with 3 lagged terms of indoor temperature, 2 of outdoor temperature, 3 of horizontal radiation, and 2 of heating requirements gave the best prediction result with a linear activation function. Ma et al. [29] used LSTM for time series prediction; thus, the future values of indoor temperature were predicted using the past measured values of parameters in the building.

In this paper, the building was modeled based on its historical measured data. Two predictive models were developed: one for forecasting indoor temperature and another for predicting gas consumption. The integration of historical data proved to be helpful when selecting the model structure. Indeed, it allowed for capturing the building's thermal inertia, thereby significantly improving the accuracy of both models. Several machine learning algorithms were tested, including multiple linear regression, random forest, artificial neural networks, LSTM, GRU models, and support vector machines. In the following sections, only the models that yielded the best performance are discussed, which are ANN and SVM.

Model training was performed on part of the data, while another part was retained to test the model. Normalization was also applied to the data. In order to evaluate the model, besides the coefficient of determination, a prediction error rate was calculated on the test dataset. The model's coefficient of determination R^2, which measures how well the observed data match those fitted, was also calculated. A coefficient of determination close to 1 indicates good model accuracy. Equation (1) shows the formula of R^2.

$$R^2 = 1 - \frac{\sum_{i=1}^{n}(y_i - \hat{y}_i)^2}{\sum_{i=1}^{n}(y_i - \overline{y})^2} \tag{1}$$

where y_i is the measured value, $\hat{y}_i$ is the predicted value, $\overline{y}$ is the mean of the observed values of y, and n is the number of observations of the test dataset.

Thus, in the presented work, three error metrics were used:

Mean Absolute Error (MAE), calculated as presented in Equation (2), yields the average of the absolute deviations from the observed values.

$$\text{MAE} = \frac{1}{n}\sum_{i=1}^{n}|y_i - \hat{y}_i| \tag{2}$$

The Root Mean Square Error (RMSE) measures the square root of the mean of the square of the errors, as shown in Equation (3):

$$\text{RMSE} = \sqrt{\frac{1}{n}\sum_{i=1}^{n}(\hat{y}_i - y_i)^2} \tag{3}$$

The Root Mean Squared Logarithmic Error (RMSLE), which is consistent with ASHRAE's evaluation method in the Kaggle competition, is mainly used when predictions have large deviations, which is the case with the energy prediction. Also, it measures the relative error between predicted and actual values while being more robust to outliers [38]. It is calculated as shown in Equation (4):

$$\text{RMSLE} = \sqrt{\frac{1}{n}\sum_{i=1}^{n}(log(\hat{y}_i + 1) - log(y_i + 1))^2} \tag{4}$$

Concerning MAE, RMSE, and RMSLE, the smaller the values of these errors are, the better the model is.

2.2.1. Artificial Neural Network for Indoor Air Temperature

The first machine learning model developed was an artificial neural network. The ANN was chosen because it has proven its efficiency in many building studies, especially in the prediction of indoor air temperature. It provided the best results in [6,37,39–41]. Also, numerical optimizations using a combination of artificial neural network and genetic algorithm can be efficient for building applications, which can save a significant amount of computational time [42].

An artificial neural network is formed from layers of interconnected artificial neurons, whose role is to find the best relation between input and output. This relation is defined by an activation function found in each neuron, and by coefficients. These coefficients are the synaptic weights of the neural network. During learning, the neural network performs several iterations to adjust the weights in a way that minimizes the error between the predicted value $\hat{y}$ and its real value y. In this paper, the ANN model was used to predict the building's indoor temperature at the next hour, denoted as T_{in}(t).

The main difficulty with neural network models is choosing the network architecture, that is, the number of layers and neurons. Several models were trained by varying the number of lagged terms n and the number of layers and neurons in the network and testing the activation function «linear», «softplus», and «sigmoid». By lagged terms, we refer to variables observed or recorded at a specific number of hours prior to the current time. Results show that the «softplus» gives the best prediction for this case study, consistently yielding the smallest error and the highest coefficient of determination compared to other activation functions in models trained with the same number of lagged terms. To avoid overloading the table, the results of ANN with «softplus» only are presented in Table 1.

Table 1. Error metrics of the trained ANN with "Softplus" activation function and different lag numbers.

Number of Lagged Terms	MAE	RMSE	R^2
0	0.14	2.27	0.14
1	0.053	0.97	0.84
2	0.031	0.6	0.93
3	0.04	0.57	0.95
4	0.022	0.42	0.97
5	0.025	0.52	0.96
6	0.028	0.55	0.95

Figure 4 shows the evolution of the MAE with varying numbers of lags, ranging from 0 (no historical building data considered) to 6 lags. Table 1 shows that the prediction error gradually decreases as the number of lagged terms increases, and remains in this trend up to 4 lagged terms. Beyond this point, the error tends to increase again. Therefore, the model that appears to be optimal for predicting indoor temperature in this study is an artificial neural network with n = 4 lags, one hidden layer with 50 neurons, and a nonlinear activation function, the «softplus» function, shown in Equation (5).

$$f(x) = log(1 + \exp(x)) \tag{5}$$

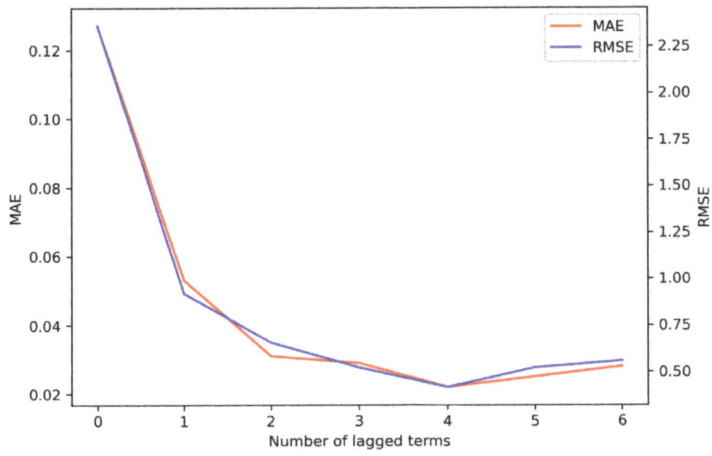

Figure 4. Evolution of MAE and RMSE errors with variations in the number of lagged terms.

The topology of the adopted model is illustrated in Figure 5. The model inputs are the historical measurements of indoor temperature, outdoor temperature, horizontal radiation, occupancy, and setpoint temperature, respectively, noted as T_{in}, T_{out}, Rad, Occ and T_{cons} from time t − 1 to t − 4. Additionally, the model relies on forecasts for outdoor temperature, solar radiation, and occupancy (at time t), as well as the setpoint temperature value proposed by the genetic algorithm T_{stp}(t). Consequently, the input layer of the neural network comprises 24 neurons: T_{in}(t − 1), T_{in}(t − 2), T_{in}(t − 3), T_{in}(t − 4), T_{out} (t − 1), T_{out}(t − 2), T_{out}(t − 3), T_{out}(t − 4), Rad(t − 1), Rad(t − 2), Rad(t − 3), Rad(t − 4), Occ(t − 1), Occ(t − 2), Occ(t − 3), Occ(t − 4), T_{stp}(t − 1), T_{stp}(t − 2), T_{stp}(t − 3), T_{stp}(t − 4), T_{out}(t), Rad(t), Occ(t), T_{stp}(t). The error measurements of the adopted model are provided in Table 2.

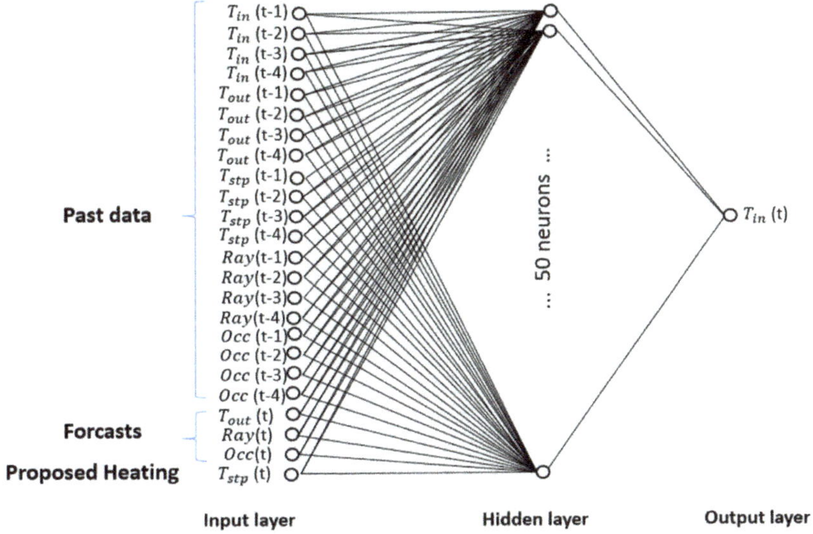

Figure 5. The adopted neural network topology.

Table 2. Error metrics for ANN model.

Model 1	R^2	RMSE	MAE
ANN	0.97	0.42	0.022

2.2.2. SVM for Gas Consumption Prediction

The second model is designed to predict the gas consumption of the boiler. As previously explained, there is no straightforward relationship between the control setpoint and gas consumption. To solve this task, several models are trained with the aim of finding the most accurate possible model, always with the available data. This raises several challenges regarding the choice of model and the choice of the number of lagged terms for any variables. The initial variables are therefore the same: T_{in}, T_{out}, Rad, Occ and T_{cons}.

In this study, random forest was used to calculate the importance of the characteristics of the model and then help in removing unnecessary variables. Random forest is a combination of decision trees. It improves the performance of a single tree by combining the randomization in the selection of partitioning data nodes in the construction of a decision tree. Hence, random forest determines feature importance by measuring each feature's contribution to reducing impurity (like Gini impurity or MSE) across all splits and trees in the ensemble, with higher reductions indicating more important features [43]. Figure 6 reveals the features' importance by a random forest with n = 2 lagged terms. It appears that taking measurements more than 1 step back does not lead to a better prediction. The results of the trained models, using different number of lags of the variables show that the best prediction can be performed using only four variables: $T_{stp}(t)$, T_{in}(t − 1), C_{gas}(t − 1), T_{stp}(t − 1). Using these features, a support vector machine (SVM) model gives the most accurate prediction in this study.

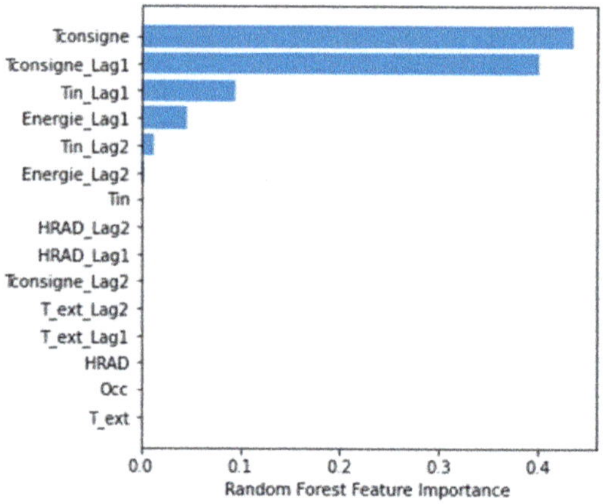

Figure 6. Importance of variables in the model with 2 lagged terms.

SVM is a supervised machine learning algorithm that aims to find an optimal boundary between possible outputs by maximizing the separation margins between data points based on their responses. This transformation of inputs is achieved using the machine's kernel function. In this study, Support Vector Regression (SVR) (a type of support vector machine) was used to analyze data for regression analysis. SVM has been commonly used for the energy prediction in buildings [30,33,36,44,45].

In this study, the radial basis function (RBF), represented in Equation (6), was chosen due to the broad and nonlinear characteristics of the dataset [46].

$$K_{RBF}(x - x') = exp\left(-\gamma x - x'^2\right) \quad (6)$$

where γ is a parameter to determine the distribution of the kernel, and $x - x'$ is the Euclidean distance between the set of points [30].

The adopted model is also based on the lag terms n = 1 of the building's indoor temperature, gas consumption, and setpoint temperature value T_{stp}(t) proposed by the genetic algorithm. The output of this model is the estimated gas consumption at time t. Among several models tested using the selected features, SVM gives the best prediction accuracy. The error measurements of the adopted model are provided in Table 3.

Table 3. Error metrics for SVM model.

Model 2	R^2	RMSLE
SVM	0.83	2.9

2.3. Optimization

As mentioned previously, the objective of this study is to identify a sequence of 24 hourly setpoint temperatures for the heating system that balances thermal comfort and energy consumption. Therefore, a multicriteria optimization problem arises. Genetic algorithms are well suited for this type of problem [47]; in particular, these algorithms have gained prominence in the optimization of building thermal systems [48].

Genetic algorithms are inspired by the process of natural selection and evolutionary biology, where an individual is represented by a chromosome composed of genes containing hereditary traits. The process begins with a population of potential solutions represented by individuals, often in the form of binary strings encoding possible solutions. In this case study, the initial population consists of setpoint temperature scenarios, and each scenario comprises 24 temperature values. Moreover, each setpoint temperature value is encoded with 6 bits, with each individual represented by 144 bits. Since each bit can take 2 possible values (0 or 1), the number of possible solutions for each proposed temperature scenario reaches more than 2.2×10^{43} solutions.

The focus of this study is on the setpoint temperature of the water supplying the radiators in the secondary circuit, rather than on optimizing the boiler's operation. Thus, the setpoint temperature values that can be chosen by the genetic algorithm range from 16 °C to 70 °C. The fitness of each solution in addressing the given problem is evaluated based on a predefined objective function. The individuals then go through genetic operations, including selection, crossover, and mutation. During selection, only the fittest solutions are chosen to become parents for the next generation. The number of generations is determined by the number of iterations. In crossover, genetic information is exchanged between parents to create offspring that exhibit a combination of their characteristics. Mutation introduces random changes in the genetic information of the offspring. This cycle of selection, crossover, and mutation repeats over multiple generations, gradually improving the overall fitness of the population and converging towards optimal solutions, or those with the best objective function scores.

In this study, the comfort score is simply estimated as a function of indoor air temperature, as in most cases [17,21,27,29]. This is because the temperature is an easily measurable parameter, compared to other parameters such as air speed or relative humidity which play a role in the notion of comfort. Therefore, ranges of comfort temperatures are defined for the genetic algorithm to adhere to. The details of these temperatures are shown in

Table 4. The objective is to maintain an indoor temperature between 20 °C and 22 °C during occupied hours (from 8 a.m. to 6 p.m. on weekdays). The comfort score of any solution that results in an indoor temperature outside this range will be penalized. Similarly, during unoccupied hours, temperatures below 15 °C are penalized. This approach aligns well with conventional control practices in real life, where similar steps are taken to prevent excessive temperature drops and the associated inefficiencies in recovery.

Table 4. Comfort temperature range during occupation/non-occupation shifts of the week.

Shift	Occupation	Target Temperature
Weekday [08 h–18 h]	Yes	[20–22 °C]
Weekday [19 h–07 h] Weekend [00 h–23 h]	No	≥15 °C

Figure 7 details the calculation of the comfort score of each hour of the day: if the building is in its occupation period (as shown in Table 4) and the indoor temperature is below the lower limit of the defined comfort temperature range, the difference between these two temperatures is added to that score. Similarly, if the indoor temperature exceeds the upper limit of the defined comfort temperature range, the difference between these two temperatures is added to the comfort score. However, if the indoor temperature is within the defined comfort temperature range, the score is not penalized. The calculation is performed similarly during non-occupation hours; if the temperature drops below 15 °C, the score is penalized.

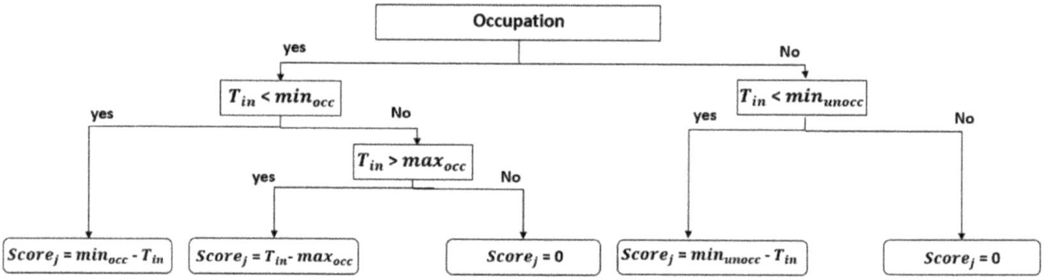

Figure 7. Diagram explaining the calculation of comfort score for each hour (j) of the day.

In this study, the comfort and gas consumption scores are normalized to facilitate balancing their weights effectively. For the comfort score calculation, an individual score is computed for each of the 24 h (as shown in Figure 7). These hourly scores are then summed up and divided by 24, as the maximum possible score for a single hour is approximately 1, resulting in a normalization factor of 24 × 1, as in Equation (7), so the value of the comfort score ranges between 0 and 1:

$$score_{comfort} = \sum_{j=1}^{24} \left(score_{comfort(j)} \right) / 24 \tag{7}$$

Concerning the consumption score, it represents the gas consumed over 24 h, as shown in Equation (8). It is normalized by dividing the summation by the maximum value of gas consumption over 24 h, so the value of the consumption score falls between 0 and 1.

$$score_{consumption} = \sum_{j=1}^{24} C_{gas(j)} / C_{gas\_max} \tag{8}$$

Since the objective of this study is to achieve a trade-off between comfort and consumption, the score for each solution is calculated by adding the score of comfort to the score of consumption. The final score is thus calculated according to Equation (9). The goal is to minimize the final score over a 24 h period.

$$score_{final} = \omega \times score_{comfort} + (1 - \omega) \times score_{consumption} \tag{9}$$

With this definition of the comfort final score ($score_{final}$), the strategy enables users to define the priority of comfort relative to energy consumption. A lower value of ω places greater emphasis on consumption, while a higher value prioritizes comfort. This way, with both sub-scores being normalized, theoretically, ω can tolerate values between 0 and 1. Thus, it becomes evident that $\omega = 1$ only considers comfort and $\omega = 0$ only considers consumption. But from a practical perspective, ω values should avoid extreme settings. Specifically, $\omega = 0$ would result in the algorithm deactivating the system, thereby eliminating comfort entirely, which is inconsistent with the intended goals of the strategy. Conversely, $\omega = 1$ would prioritize comfort to the exclusion of energy efficiency, potentially leading to excessive energy consumption, which similarly contradicts the strategy's overarching objectives.

The choice of the value of ω is left to the discretion of the building manager. The weight ω specifies the priority level of comfort relative to consumption, meaning ω can be adjusted to favor one score over the other. Since the comfort score and the energy consumption score are qualitatively different metrics, they cannot be rigorously balanced against one another. In this study, the authors selected a value of $\omega = 0.5$, a midpoint of the [0, 1] interval, to serve as a neutral baseline, in order to combine the two criteria in one objective function. We proceeded under the hypothesis that this value aligns with the overarching aim of the strategy to optimize system performance while maintaining a practical equilibrium between the competing priorities. Therefore, the results presented in this article were obtained with a weight of $\omega = 0.5$.

As explained, the optimization using the genetic algorithm is initiated every hour in an infinite loop (loop 3 of the methodology as explained in Section 2), enabling a receding horizon. Each iteration of this loop represents the calculation of an optimal sequence for 24 h. At time t_0, the GA finds an optimal solution, adhering to the defined number of solutions for the initial population and the number of iterations to be performed. In this study, we focused on reducing the computational complexity by helping the genetic algorithm find the best solution at time $t_0 + 1$ by injecting the best solution found at the first optimization (t_0). The process is explained in Figure 8. During the second iteration, i.e., at time $t_0 + 1$, the GA can use the previous solution, including it in the new initial population of 20 solutions instead of 50. In addition, the GA performs 15 iterations instead of 25, thereby reducing execution time while maintaining the efficiency of the process (more details are presented in Section 3). Therefore, the number of sequences is limited to 50 for the first iteration of the receding horizon loop (loop 3), and then reduced to 20 for the remaining 47 iterations in this case study (since the loop turns for 48 h).

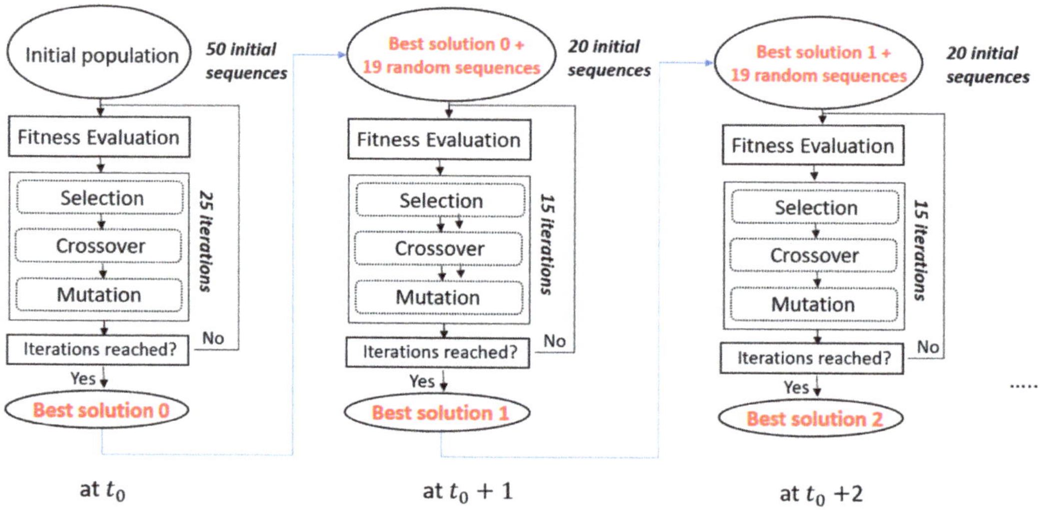

Figure 8. Explanation of the genetic algorithm generation loop.

3. Results

In this study, the first loop of the proposed MPC represents the prediction of indoor temperature and gas consumption over a 24 h horizon, while the second loop denotes the optimization process of the setpoint temperature. Concerning the third one, it represents the receding horizon where the prediction and the optimization are performed at each timestep (every hour), for the next 24 h, in order to adjust the heating scenario if needed. In real-world scenarios, the third loop operates continuously from the moment the strategy is activated until it is deactivated. This means that the forecasting horizon advances incrementally on an hourly basis throughout the duration of the strategy's implementation. However, in this study, to reduce computational effort and time, the authors chose to validate the strategy by running it for a fixed duration of 48 h loops, applied to 10 randomly selected days. The computation of the optimal solution for one timestep, if the loop can range from 10 to 40 min, is carried out using a relatively high-performance machine (DELL PRECISION 3581, with a 13th Gen Intel Core i9-13900H processor, 64 GB of RAM, and a 64-bit architecture), making it challenging to perform large-scale calculations, such as those spanning an entire year.

The proposed MPC strategy was validated on the modeled building using TRNSYS18. Figure 9 explains the mechanism of this loop. At each iteration, the first setpoint temperature (T_{stp}(t)) of the proposed 24 h scenario is applied to the heating system of the simulated building model in TRNSYS18 as a control signal. This allows for the collection of output data, including the indoor air temperature and gas consumption associated with the specified heating setpoint. The simulation is then executed using the building's historical data, maintaining consistent initial conditions for both the prediction process and the simulation, up to the considered timestep. Once the simulation on TRNSYS18 is performed, the output values of indoor temperature and energy consumption (T_{in}(t) and C_{gas}(t), respectively) are passed to update the dataset used for the prediction and optimization processes at the next timestep. This study was conducted in hourly steps, i.e., one timestep is equal to one hour.

As explained in Section 2, at each timestep, the genetic algorithm attempts to find a solution that best accommodates changes or uncertainties in weather and occupancy. Figure 10 shows an example of evolution of the heating scenarios during three consecutive iterations (from the timestep t at 5 a.m. (scenario t), to scenario t + 1 at 6 a.m., and scenario t + 2 at 7 a.m.). The heating scenario dynamically adjusts to explore improved solutions that enhance comfort while incorporating updated values. This can be more explicit in real-life application, where potential weather or occupancy changes may occur. Figure 11 shows the indoor temperature that results from each heating scenario presented in Figure 10. The adjustment of the scenario at t + 2 raises the indoor temperature a bit more to try to reach the comfort range temperatures during the occupancy period.

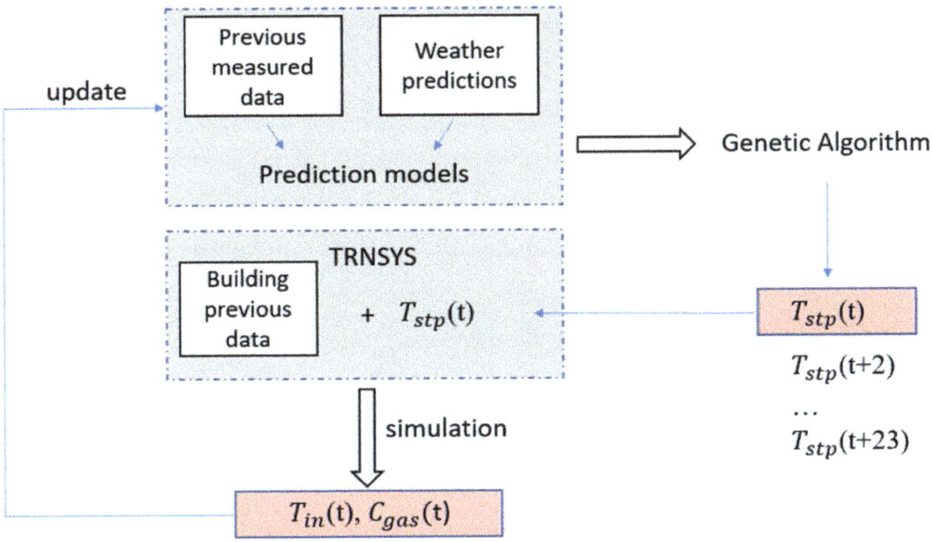

Figure 9. Validation process via TRNSYS18: one iteration of the receding horizon loop of the MPC.

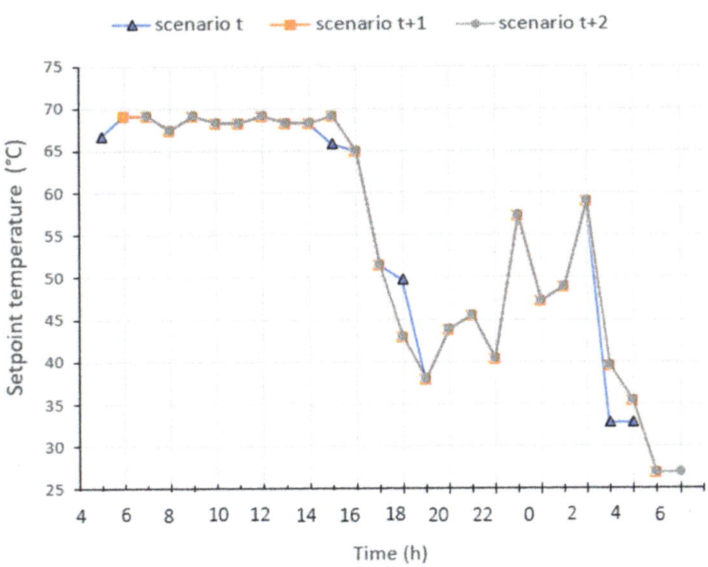

Figure 10. An example of the evolution of the heating scenario over the receding horizon.

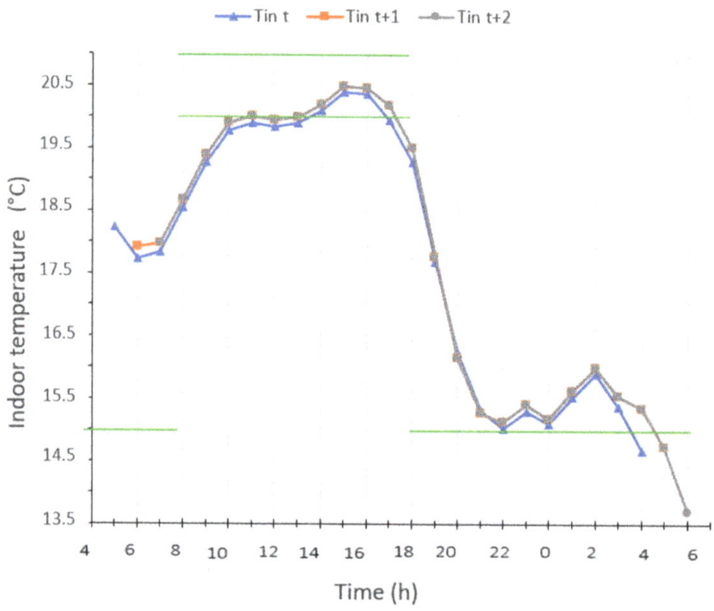

Figure 11. The evolution of the indoor temperature resulting from the change of the heating scenario of Figure 10. The green lines mark the temperature constraints of comfort during occupation and unoccupancy of the building.

In order to validate the performance of the strategy, 10 random days were selected throughout the heating period. It is important to clarify that these are independent tests. For each of these days, the implementation of the MPC strategy was initiated 24 h in advance to avoid the transition phase between the conventional control strategy and the predictive control strategy. The changes in indoor temperature and gas consumption over the last 24 h of each period were used to calculate the scores. The calculation was thus launched in a 48 h receding horizon loop. That means that 48 iterations of loop 3 were performed for every selected day. The optimization starts at the first timestep t, calculating the 24 values of setpoint temperatures for the next 24 h, considering the next 24 h occupation scenario and weather forecasts. Then, at the next timestep, the calculation of the heating scenario is repeated for the 24 next ones, and so on, for 48 iterations. Figure 12 shows an example of indoor temperature and gas consumption that result from the MPC strategy applied to 1 selected test day, over 48 h. Only the values of the second day were used to evaluate the applied strategy (the part selected in the grey frame on the graph).

Values of indoor air temperature and energy consumption from the last 24 h were compared to those simulated using a conventional control strategy for the heating system of the same building. This conventional control operates based on a heating curve, a function of the outdoor temperature. The parameters of this heating curve were carefully chosen to ensure an indoor temperature between 20 °C and 22 °C during occupation hours (8 a.m. to 6 p.m.). The heating setpoint temperature starts to increase at 6 a.m. (2 h before the occupation to guarantee a comfort temperature at 8 a.m.), with a night-time setback also implemented. The heating curve was optimally adjusted to provide a credible reference strategy. In other words, the baseline method with which we compared our strategy is already a fairly energy-efficient heating curve, and a heating cutdown is carried out during the night. Figure 13 shows the indoor air temperature in the building, with heating based on the selected heating curve over the course of a week. It is clear that it takes hours to reach a comfortable temperature. The proposed MPC aims to avoid this delay. The results presented in this section correspond to the comparison of the strategy with this selected conventional heating strategy.

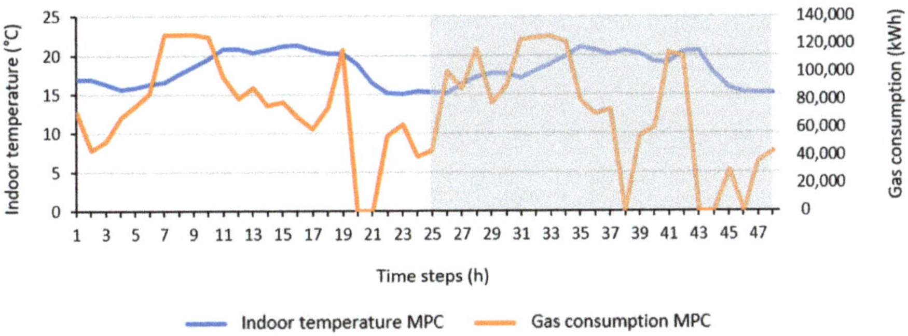

Figure 12. Indoor temperature and gas consumption that result from the simulation considering the values of the receding horizon heating scenario over 48 timesteps. The values of the last 24 h are used for validation (grey frame).

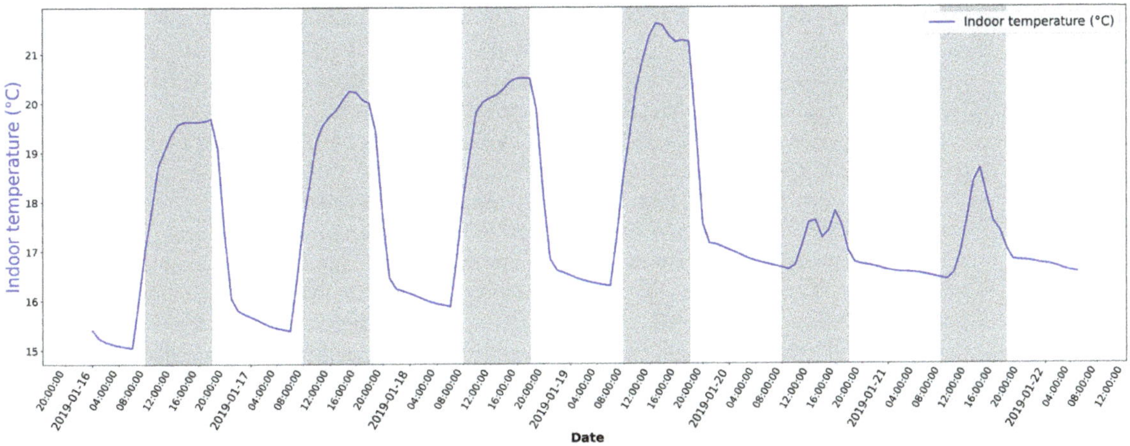

Figure 13. Indoor air temperature curve with the selected heating curve strategy, over a week. The grey blocks mark the occupation period of the building (8 h–18 h).

Since the aim of the strategy is to balance energy consumption and thermal comfort, the comparison is based on two scores: a comfort score and an energy consumption score. The MPC strategy aims to minimize the final score, which is the sum of both scores, as defined in Equation (9). Table 5 shows the results of both control methods over ten randomly selected days, representing different weather conditions (varying outdoor temperatures, solar radiation). The control using the baseline heating curve is abbreviated as RBC (Rule-Based Control). The calculation is performed over 48 hourly steps for each day as mentioned, meaning that during these 48 h, the optimization is reinitiated every hour with a 24 h horizon each time. The scores presented in Table 5 are calculated on the second day (the last 24 h) to ensure that the strategy is evaluated based on a steady-state regime.

Table 5. MPC vs. RBC for the 10 days tested.

Day	Average Solar Radiation (W/m²)	Average Outdoor Temperature (°C)	Comfort Score MPC	Comfort Score RBC	Consumption Score MPC	Consumption Score RBC	Final Score MPC	Final Score RBC	Energy Savings (%)
1	76.2	3.3	0.18	0.18	0.76	0.81	0.47	0.5	6.2
2	34.1	8.2	0.14	0.15	0.4	0.57	0.27	0.36	29.8
3	20.6	−0.6	0.16	0.15	0.62	0.64	0.39	0.39	3.1
4	27.4	7	0.27	0.18	0.39	0.55	0.33	0.36	29
5	23.9	3	0.17	0.14	0.64	0.71	0.4	0.42	9.9
6	38.2	0	0.11	0.13	0.73	0.78	0.42	0.45	6.4
7	5.4	3.3	0.23	0.13	0.65	0.7	0.44	0.42	7.1
8	40.7	3.1	0.37	0.14	0.61	0.69	0.49	0.41	11.6
9	85.1	3.7	0.12	0.29	0.54	0.67	0.33	0.47	19.4
10	31.7	0	0.14	0.15	0.72	0.78	0.43	0.46	7.7

Table 5 shows the comfort scores of both strategies in every selected test day. We can see that, in almost all cases, the MPC comfort scores are equal or lower than the RBC scores (days 1, 2, 6, 9, and 10), while an energy consumption reduction is proved by the MPC strategy in all presented days. Here, it is worth mentioning that, since there is a comfort range of temperature that is defined for the occupation and the unoccupancy period as well, the comfort score could be penalized for the violation of both. That means that, even during unoccupancy, if the temperature drops below 15 °C (Table 4), the comfort score will increase; however, this does not imply that the building occupants feel discomfort.

Figures 14 and 15 compare, respectively, the indoor temperature and the energy consumption of both strategies on day 4 presented in Table 5. The MPC strategy was able to anticipate the solar radiation expected during the day, to keep the indoor temperature within the comfort range (between 20 °C and 22 °C) during occupancy hours (grey frame). The results show lower gas consumption in the case of the MPC strategy, compared to that of the RBC where the indoor temperature exceeded the defined comfort range (above 22 °C). The comparison of the final scores over the 10 selected days is visualized in Figure 16. It shows that the MPC strategy outperforms the conventional strategy in most cases (lower final scores for MPC), except for days 7 and 8. This is due to the MPC comfort score which is higher than that of the RBC during these two days. Figure 17 compares the values of indoor temperature corresponding to the MPC against those of the RBC strategy over the 24 h timesteps of day 8. The reason behind the higher comfort score of the MPC is the decrease in temperatures during unoccupancy, as already explained. This cutdown of heating did not prohibit the strategy of restoring the comfort temperatures during occupation hours. Results show that in day 8, the MPC strategy consumed 11.6% less energy compared to the RBC strategy. This demonstrates that the proposed strategy was able to anticipate the expected solar radiation for the day and suggested heating setpoints that maintain comfort temperatures while minimizing energy consumption. Figure 18 compares the gas consumption of both strategies for day 8.

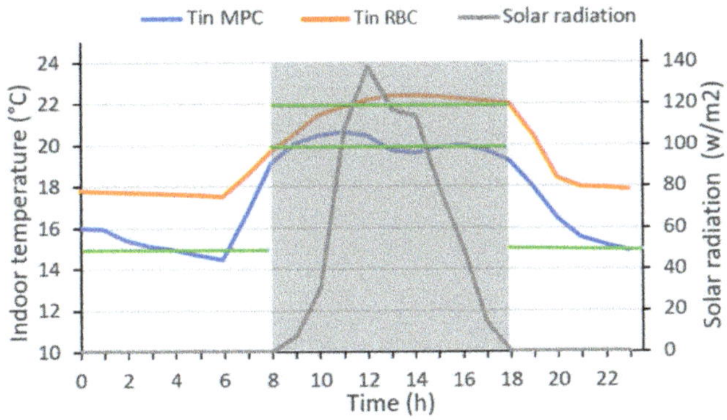

Figure 14. Indoor temperature of both strategies: MPC (blue curve) and RBC (orange curve) on day 4, while the curve represents the solar radiation during the day. The grey frame denotes occupation shift, while the green lines mark the temperature constraints of comfort during occupation and unoccupancy of the building.

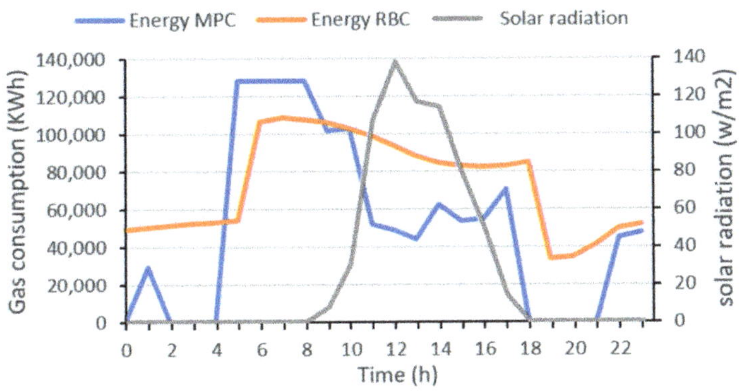

Figure 15. Gas consumption of both strategies: MPC (blue curve) and RBC (orange curve) on day 4. The grey curve represents the solar radiation.

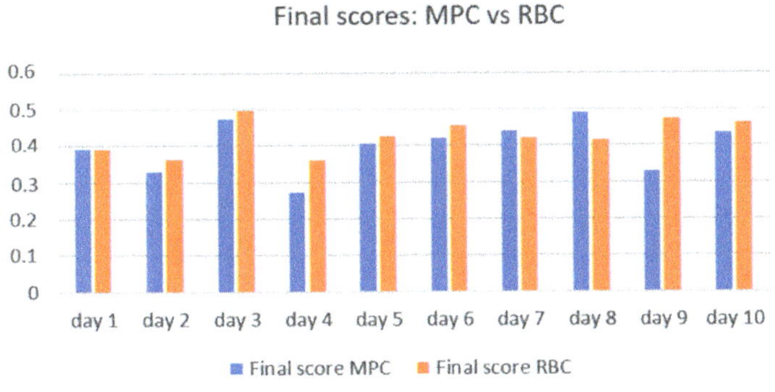

Figure 16. Comparison of final scores.

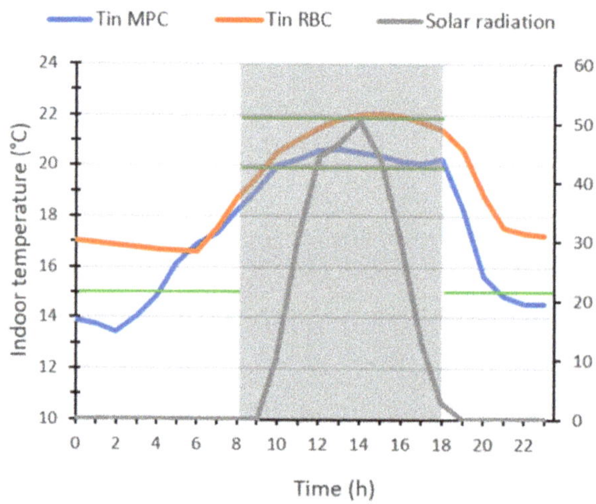

Figure 17. Indoor temperature of both strategies: MPC (blue curve) and RBC (orange curve) on day 8. The grey curve represents the solar radiation, and grey frame represents the occupation, while the green lines mark the temperature constraints of comfort during occupancy of the building.

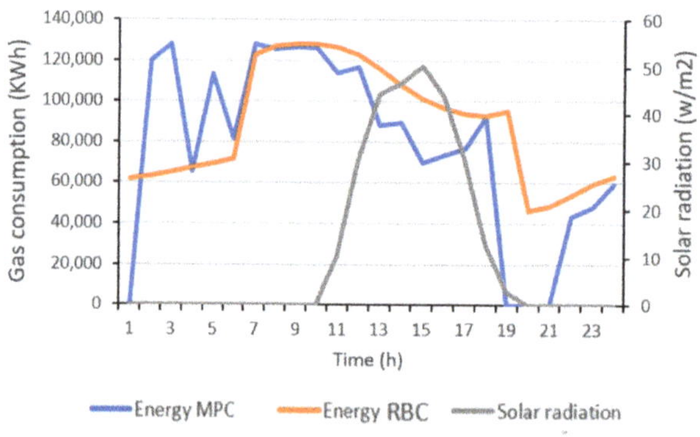

Figure 18. Gas consumption of both strategies: MPC (blue curve) and RBC (orange curve) on day 8. The grey curve represents the solar radiation.

4. Discussion

In this paper, a predictive control strategy for a building's heating system with radiators powered by a gas boiler was proposed aiming to balance thermal comfort and gas consumption. For this study, the occupant's comfort was represented by the indoor air temperature, a common criterion in current building regulation systems. While future work may consider factors such as indoor air relative humidity, the study focused solely on temperature as the comfort indicator.

The proposed method relies on two prediction models informed by conventional building data, integrating weather and occupancy forecasts. The selection of these models was discussed in the paper, where several machine learning models were tested, and those demonstrating the highest prediction accuracy were chosen. The integration of historical data significantly improved precision. The neural network model aimed at predicting indoor temperature proved to be highly accurate, but the support vector machine model selected for predicting gas consumption, crucial for implementing the method, appears to have potential for improvement. The decision not to rely on internal measurements of technical equipment, such as various temperatures from the primary and secondary parts of the heating system, adds complexity to the modeling process. This aspect could be the focus of specific future research efforts. The two models, used to predict temperature and consumption over a 24 h forecasting horizon, are integrated within an optimization loop using genetic algorithms. The optimization phase aims to define an optimal sequence of setpoint temperatures for the secondary circuit's water supply in the heating system on an hourly basis. The process of prediction and the calculation of the 24 h sequence of heating setpoint temperatures is repeated every hour, replacing the predicted values of indoor temperature and gas consumption at the next timestep, by the values simulated on the modeled building, preventing the accumulation of prediction errors throughout the horizon.

The method was developed using data generated in TRNSYS18. A case study building was modeled, and the simulation tool was employed to test and compare the predictive control strategy against a conventional Rule-Based Control strategy typically used in such buildings. It has been confirmed that the proposed MPC strategy can anticipate potential events (weather disturbances or internal gains) that could disrupt the building's behavior, while achieving energy savings and maintaining occupant thermal comfort.

Because MPC strategies are implemented in buildings that vary in heating systems, environmental and weather conditions, and initial operating states, directly comparing their performance across studies is inherently challenging. Instead, comparison was performed against a reference model tailored to the same building, departing from the same initial conditions. By comparing our MPC-based strategy with an already optimized heating curve, we demonstrated additional energy savings, ranging between 3% and 30% across various randomly tested days. These results were obtained by combining a comfort score and a consumption score for each of these test days. A weighting between the two scores had to be chosen to achieve a compromise between thermal comfort and energy consumption. This weighting determines the balance of this compromise, and the choice of its value is left to the discretion of the operators, based on their priority for each part of the objective function.

This study shows that an MPC strategy can indeed be deployed with only a small set of readily obtained measurements, enabling straightforward transfer to other buildings with similar heating systems. Moving forward with the research, the goal is to implement the proposed control method on a real case study building. Measurements are currently being collected, and a predictive strategy is expected to be developed.

Author Contributions: Conceptualization, R.L.; Methodology, R.L.; Software, J.C.; Validation, R.L. and D.D.; Formal analysis, R.L.; Resources, R.L.; Data curation, R.L.; Writing—original draft, R.L.; Writing—review & editing, D.D., O.A.-H. and J.C.; Supervision, D.D.; Project administration, D.D. All authors have read and agreed to the published version of the manuscript.

Funding: This research received no external funding.

Data Availability Statement: The original contributions presented in the study are included in the article, further inquiries can be directed to the corresponding author/s.

Acknowledgments: This work was performed with help of Artois University and the Hauts-de-France Region.

Conflicts of Interest: The authors declare no conflicts of interest. The funders had no role in the design of the study; in the collection, analyses, or interpretation of data; in the writing of the manuscript; or in the decision to publish the results.

Abbreviations

The following abbreviations are used in this manuscript:

Nomenclature	
T_{stp}	Setpoint temperature
T_{in}	Indoor temperature
T_{out}	Outdoor temperature
C_{gas}	Gas consumption
Occ	Occupation
Rad	Horizontal solar radiation
min_{occ}	Lower limit of comfort temperature range during occupation
max_{occ}	Upper limit of comfort temperature range during occupation
min_{unocc}	Lower limit of allowed temperature range during unoccupancy
MPC	Model Predictive Control
RC	Resistance–Capacitance
3W valve	3-way valve
HVAC	Heating, ventilation, air-conditioning
ANN	Artificial neural network
SVR	Support Vector Regressor
HDD	Heating degree days
PMV	Predicted Mean Vote
SVM	Support vector machine
RBF	Radial basis function
ARX	Autoregressive model with exogenous inputs
MAE	Mean Average Error
RMSE	Root Mean Square Error
GA	Genetic algorithm
R^2	Coefficient of determination
RMSLE	Root Mean Square Log Error

References

1. Santamouris, M.; Vasilakopoulou, K. Present and Future Energy Consumption of Buildings: Challenges and Opportunities towards Decarbonisation. *e-Prime Adv. Electr. Eng. Electron. Energy* **2021**, *1*, 100002. [CrossRef]
2. Khadra, A.; Hugosson, M.; Akander, J.; Myhren, J.A. Development of a Weight Factor Method for Sustainability Decisions in Building Renovation. Case Study Using Renobuild. *Sustainability* **2020**, *12*, 7194. [CrossRef]
3. Pinzon Amorocho, J.A.; Hartmann, T. A multi-criteria decision-making framework for residential building renovation using pairwise comparison and TOPSIS methods. *J. Build. Eng.* **2022**, *53*, 104596. [CrossRef]
4. Suszanowicz, D. Optimisation of Heat Loss through Ventilation for Residential Buildings. *Atmosphere* **2018**, *9*, 95. [CrossRef]
5. Zhou, L.; Haghighat, F. Optimization of ventilation system design and operation in office environment, Part I: Methodology. *Build. Environ.* **2009**, *44*, 651–656. [CrossRef]
6. Yang, S.; Wan, M.P.; Chen, W.; Ng, B.F.; Dubey, S. Model predictive control with adaptive machine-learning-based model for building energy efficiency and comfort optimization. *Appl. Energy* **2020**, *271*, 115147. [CrossRef]
7. Chow, T.T.; Zhang, G.Q.; Lin, Z.; Song, C.L. Global optimization of absorption chiller system by genetic algorithm and neural network. *Energy Build.* **2002**, *34*, 103–109. [CrossRef]
8. Li, Q.; Meng, Q.; Cai, J.; Yoshino, H.; Mochida, A. Applying support vector machine to predict hourly cooling load in the building. *Appl. Energy* **2009**, *86*, 2249–2256. [CrossRef]
9. Rijksen, D.O.; Wisse, C.J.; van Schijndel, A.W.M. Reducing peak requirements for cooling by using thermally activated building systems. *Energy Build.* **2010**, *42*, 298–304. [CrossRef]
10. Pérez-Lombard, L.; Ortiz, J.; Pout, C. A review on buildings energy consumption information. *Energy Build.* **2008**, *40*, 394–398. [CrossRef]
11. Xin, X.; Zhang, Z.; Zhou, Y.; Liu, Y.; Wang, D.; Nan, S. A comprehensive review of predictive control strategies in heating, ventilation, and air-conditioning (HVAC): Model-free vs. model. *J. Build. Eng.* **2024**, *94*, 110013. [CrossRef]
12. Li, H.; Wang, S. Comparative assessment of alternative MPC strategies using real meteorological data and their enhancement for optimal utilization of flexibility-resources in buildings. *Energy* **2022**, *244*, 122693. [CrossRef]

13. Avci, M.; Erkoc, M.; Rahmani, A.; Asfour, S. Model predictive HVAC load control in buildings using real-time electricity pricing. *Energy Build.* **2013**, *60*, 199–209. [CrossRef]
14. Mohsenian-Rad, A.-H.; Leon-Garcia, A. Optimal Residential Load Control With Price Prediction in Real-Time Electricity Pricing Environments. *IEEE Trans. Smart Grid* **2010**, *1*, 120–133. [CrossRef]
15. Bamdad, K.; Mohammadzadeh, N.; Cholette, M.; Perera, S. Model Predictive Control for Energy Optimization of HVAC Systems Using EnergyPlus and ACO Algorithm. *Buildings* **2023**, *13*, 3084. [CrossRef]
16. Ferreira, P.M.; Ruano, A.E.; Silva, S.; Conceição, E.Z.E. Neural networks based predictive control for thermal comfort and energy savings in public buildings. *Energy Build.* **2012**, *55*, 238–251. [CrossRef]
17. Mylonas, A.; Macià-Cid, J.; Péan, T.Q.; Grigoropoulos, N.; Christou, I.T.; Pascual, J.; Salom, J. Optimizing Energy Efficiency with a Cloud-Based Model Predictive Control: A Case Study of a Multi-Family Building. *Energies* **2024**, *17*, 5113. [CrossRef]
18. Abdellatif, M.; Chamoin, J.; Nianga, J.-M.; Defer, D. A thermal control methodology based on a machine learning forecasting model for indoor heating. *Energy Build.* **2022**, *255*, 111692. [CrossRef]
19. Reynolds, J.; Rezgui, Y.; Kwan, A.; Piriou, S. A zone-level, building energy optimisation combining an artificial neural network, a genetic algorithm, and model predictive control. *Energy* **2018**, *151*, 729–739. [CrossRef]
20. Peeters, L.; Van Der Veken, J.; Hens, H.; Helsen, L.; D'haeseleer, W. Control of heating systems in residential buildings: Current practice. *Energy Build.* **2008**, *40*, 1446–1455. [CrossRef]
21. Aoun, N.; Bavière, R.; Vallée, M.; Aurousseau, A.; Sandou, G. Modelling and flexible predictive control of buildings space-heating demand in district heating systems. *Energy* **2019**, *188*, 116042. [CrossRef]
22. Chen, W.-H.; You, F. Sustainable building climate control with renewable energy sources using nonlinear model predictive control. *Renew. Sustain. Energy Rev.* **2022**, *168*, 112830. [CrossRef]
23. Váňa, Z.; Cigler, J.; Široký, J.; Žáčeková, E.; Ferkl, L. Model-based energy efficient control applied to an office building. *J. Process Control* **2014**, *24*, 790–797. [CrossRef]
24. Fielsch, S.; Grunert, T.; Stursberg, M.; Kummert, A. Model Predictive Control for Hydronic Heating Systems in Residential Buildings. *IFAC-Pap.* **2017**, *50*, 4216–4221. [CrossRef]
25. Široký, J.; Oldewurtel, F.; Cigler, J.; Prívara, S. Experimental analysis of model predictive control for an energy efficient building heating system. *Appl. Energy* **2011**, *88*, 3079–3087. [CrossRef]
26. Prívara, S.; Široký, J.; Ferkl, L.; Cigler, J. Model predictive control of a building heating system: The first experience. *Energy Build.* **2011**, *43*, 564–572. [CrossRef]
27. Macarulla, M.; Casals, M.; Forcada, N.; Gangolells, M. Implementation of predictive control in a commercial building energy management system using neural networks. *Energy Build.* **2017**, *151*, 511–519. [CrossRef]
28. Si, Q.; Wei, J.; Li, Y.; Cai, H. Optimization for the Model Predictive Control of Building HVAC System and Experimental Verification. *Buildings* **2022**, *12*, 1602. [CrossRef]
29. Ma, L.; Huang, Y.; Zhang, J.; Zhao, T. A Model Predictive Control for Heat Supply at Building Thermal Inlet Based on Data-Driven Model. *Buildings* **2022**, *12*, 1879. [CrossRef]
30. Shapi, M.K.M.; Ramli, N.A.; Awalin, L.J. Energy consumption prediction by using machine learning for smart building: Case study in Malaysia. *Dev. Built Environ.* **2021**, *5*, 100037. [CrossRef]
31. Wang, Z.; Wang, Y.; Srinivasan, R.S. A novel ensemble learning approach to support building energy use prediction. *Energy Build.* **2018**, *159*, 109–122. [CrossRef]
32. Smarra, F.; Jain, A.; de Rubeis, T.; Ambrosini, D.; D'Innocenzo, A.; Mangharam, R. Data-driven model predictive control using random forests for building energy optimization and climate control. *Appl. Energy* **2018**, *226*, 1252–1272. [CrossRef]
33. Zhao, J.; Li, J.; Shan, Y. Research on a forecasted load-and time delay-based model predictive control (MPC) district energy system model. *Energy Build.* **2021**, *231*, 110631. [CrossRef]
34. Afram, A.; Janabi-Sharifi, F.; Fung, A.S.; Raahemifar, K. Artificial neural network (ANN) based model predictive control (MPC) and optimization of HVAC systems: A state of the art review and case study of a residential HVAC system. *Energy Build.* **2017**, *141*, 96–113. [CrossRef]
35. Lazarevic, S.; Congradac, V.; Andjelkovic, A.; Kljajic, M.; Kanovic, Z. District heating substation elements modeling for the development of the real-time model. *Therm. Sci.* **2019**, *23*, 2061–2070. [CrossRef]
36. Fu, Y.; Li, Z.; Zhang, H.; Xu, P. Using Support Vector Machine to Predict Next Day Electricity Load of Public Buildings with Sub-metering Devices. *Procedia Eng.* **2015**, *121*, 1016–1022. [CrossRef]
37. Mechaqrane, A.; Zouak, M. A comparison of linear and neural network ARX models applied to a prediction of the indoor temperature of a building. *Neural Comput. Appl.* **2004**, *13*, 32–37. [CrossRef]
38. Fu, C.; Miller, C. Using Google Trends as a proxy for occupant behavior to predict building energy consumption. *Appl. Energy* **2022**, *310*, 118343. [CrossRef]
39. Potočnik, P.; Vidrih, B.; Kitanovski, A.; Govekar, E. Neural network, ARX, and extreme learning machine models for the short-term prediction of temperature in buildings. *Build. Simul.* **2019**, *12*, 1077–1093. [CrossRef]

40. Garnier, A.; Eynard, J.; Caussanel, M.; Grieu, S. Predictive control of multizone heating, ventilation and air-conditioning systems in non-residential buildings. *Appl. Soft Comput.* **2015**, *37*, 847–862. [CrossRef]
41. Huang, H.; Chen, L.; Hu, E. A neural network-based multi-zone modelling approach for predictive control system design in commercial buildings. *Energy Build.* **2015**, *97*, 86–97. [CrossRef]
42. Magnier, L.; Haghighat, F. Multiobjective optimization of building design using TRNSYS simulations, genetic algorithm, and Artificial Neural Network. *Build. Environ.* **2010**, *45*, 739–746. [CrossRef]
43. Hasan, M.A.M.; Nasser, M.; Ahmad, S.; Molla, K.I. Feature Selection for Intrusion Detection Using Random Forest. *J. Inf. Secur.* **2016**, *7*, 129–140. [CrossRef]
44. Li, Q.; Meng, Q.; Cai, J.; Yoshino, H.; Mochida, A. Predicting hourly cooling load in the building: A comparison of support vector machine and different artificial neural networks. *Energy Convers. Manag.* **2009**, *50*, 90–96. [CrossRef]
45. Jain, A.; Satish, B. Clustering based Short Term Load Forecasting using Support Vector Machines. In Proceedings of the 2009 IEEE Bucharest PowerTech, Bucharest, Romania, 28 June–2 July 2009; pp. 1–8.
46. Ben-Hur, A.; Ong, C.S.; Sonnenburg, S.; Schölkopf, B.; Rätsch, G. Support Vector Machines and Kernels for Computational Biology. *PLoS Comput. Biol.* **2008**, *4*, e1000173. [CrossRef] [PubMed]
47. Konak, A.; Coit, D.W.; Smith, A.E. Multi-objective optimization using genetic algorithms: A tutorial. *Reliab. Eng. Syst. Saf.* **2006**, *91*, 992–1007. [CrossRef]
48. Wright, J.A.; Loosemore, H.A.; Farmani, R. Optimization of building thermal design and control by multi-criterion genetic algorithm. *Energy Build.* **2002**, *34*, 959–972. [CrossRef]

energies

MDPI

Article

Smart Thermostat Development and Validation on an Environmental Chamber Using Surrogate Modelling

Leonidas Zouloumis [1], Nikolaos Ploskas [2], Nikolaos Taousanidis [1] and Giorgos Panaras [1,*]

[1] Department of Mechanical Engineering, University of Western Macedonia, 50132 Kozani, Greece; l.zouloumis@uowm.gr (L.Z.); ntaousanidis@uowm.gr (N.T.)
[2] Department of Electrical & Computer Engineering, University of Western Macedonia, 50132 Kozani, Greece; nploskas@uowm.gr
* Correspondence: gpanaras@uowm.gr; Tel.: +30-246-105-6644

Abstract

The significant contribution of buildings to the global primary energy consumption necessitates the application of energy management methodologies at a building scale. Although dynamic simulation tools and decision-making algorithms are core components of energy management methodologies, they are often accompanied by excessive computational cost. As future controlling structures tend to become autonomized in building heating layouts, encouraging distributed heating services, the research scope calls for creating lightweight building energy system modeling as well monitoring and controlling methods. Following this notion, the proposed methodology turns a programmable controller into a smart thermostat that utilizes surrogate modeling formed by the ALAMO approach and is applied in a 4-m-by-4-m-by-2.85-m environmental chamber setup heated by a heat pump. The results indicate that the smart thermostat trained on the indoor environmental conditions of the chamber for a one-week period attained a predictive *RMSE* of 0.082–0.116 °C. Consequently, it preplans the heating hours and applies preheating controlling strategies in real time effectively, using only the computational power of a conventional controller, essentially managing to attain at least 97% thermal comfort on the test days. Finally, the methodology has the potential to meet the requirements of future building energy systems featured in urban-scale RES-based district heating networks.

Keywords: smart thermostat; surrogate models; thermal comfort; control strategies; heat pumps; building energy management systems

1. Introduction

1.1. Building Energy Management

The scientific community has recognized the contribution of the building energy demand to the EU's primary energy consumption [1], which amounts to approximately 40%. Furthermore, the same holds true for the U.S. and Asian building sectors as well [2,3], making approaches on building thermal load management optimization a highly sought out subject. At the same time, actions are being taken to achieve energy autonomy by utilizing renewable energy sources (RESs) [4,5] but also energy demand reduction in the building sector [6,7]. Since building load optimization is such a broad subject, the research community tends to divide it into two distinct categories that have a synergetic link between them, namely demand-side management and supply-side energy management [8,9]. Also, the majority of the end-use of the energy requested by buildings concerns thermal applications, such as heating, cooling, and ventilation.

 https://doi.org/10.3390/en18133433

In a short analysis of supply-side management, this category concerns the optimal operation of energy producers. Examples regarding the implementation of supply-side management methodologies are spotted in several works. For example, ref. [10] have elaborated on various methodologies that include the use of multi-energy systems for energy production and storage, both on a thermal and electrical energy basis. The methodology of [11] achieves the reduction in demanded loads and by extent the minimization of fuel consumption, assisting the energy producers to increase their energy production efficiency and therefore reduce their operational costs. It is observed that the optimization criteria of supply-side management that concern the producers usually relate to their environmental footprint and their operational cost. Moreover, the actions regarding supply-side management are summarized on "when and how" the energy loads have to be produced and supplied to the consumers. At the same time, those actions also extend to the provision of incentives to the consumers, to encourage them to alter their own optimization criteria objectives, which are discussed below, and consequently achieve a "silver lining" between the criteria of the producers and the consumers. The most common incentivization is variable energy tariffs, which incentivize the consumers to request their loads when energy production prices are low [12]. The configuration of those energy tariff profiles depends on the expected environmental footprint and/or energy production cost reduction. It is also emphasized that incentives reach their full potential only when the consumers have the necessary knowledge of the incentive mechanism and use it to suit their best interests.

The complementary aspect of supply-side energy management, the optimal management of demand-side energy loads, is related to the actions that consumers take to satisfy their own needs posed as optimization criteria. More specifically, approaches from previously mentioned works [10,11] utilize the energy storage potential of the end-user systems to be provided to the consumers when needed. Furthermore, in the thermal flexibility enhancement methodology presented by [13], it was observed that the utilization of thermal energy storage and consumer thermal comfort levels significantly encourage RES participation in the coverage of the demanded energy loads of end-users. Summarizing demand-side management, the optimal management of energy loads aims to maximize the thermal comfort levels of the consumers during building occupancy periods while also maintaining the cost of energy used at the lowest levels. The optimization actions of the consumers include the weighing of the optimization criteria between each other, i.e., the thermal comfort and cost of requested energy, as well as the forming of thermal load profile formation, according to those criteria. As already mentioned, the criteria in the weighting process can also be partly influenced by the supply side, namely the energy producers. Consequently, the joint optimization of the demand side and supply side is heavily dependent on a coordination system that supports bidirectional communication between consumers and producers. This is even more crucial in district heating network (DHN) layouts, where the producer and the consumer are individual entities with their own different optimization criteria, which often contradict each other.

The value of utilizing energy flexibility emerges from the above discussion. On a thermal energy basis, thermal flexibility is defined as the ability of the system to alter their thermal energy injection/extraction rate in a system [14]. The primary parameters that affect the thermal flexibility of a system are the thermal storage potential, as well as the thermal power variability enabled by operational control strategies [14,15]. However, thermal comfort should also not be neglected, as there are indications from works that it is also an important parameter to thermal flexibility due to its dynamic nature over the course of a daily system operation, as well as its fluctuating behavior [16]. It is worth mentioning that shifting the occupant thermal comfort zone by 1 °C (reduction during heating operation and increase during cooling operation) can lead to substantial energy savings and a reduction in thermal peak loads [17,18].

1.2. Utilizing Thermal Flexibility with Smart Controlling

At this point, emphasis is placed on the utilization of thermal flexibility for the demand-side energy management of a building. According to the building thermal flexibility analysis above, for its maximum exploitation towards attaining desired thermal comfort settings combined with minimized energy consumption, a control structure is required with the following characteristics:

- Knowledge of the available thermal storage capacity;
- Knowledge of the desired thermal comfort zone of the user/consumer;
- Knowledge of the cost of the primary energy consumption related to the available energy sources;
- Ability to schedule operation controlling strategies of the heating systems of the building to enable power variability.

For the effective use of the four above characteristics of the controlling structure, the ability to predict the future states of the building is crucial. In other words, a dynamic simulation model of the building layout should be available. However, the local installation of such simulation models in building infrastructures becomes more difficult due to the system requirements in terms of computational power, and by extent, investment and operational cost [19]. This is something that should be considered when installing smart controlling layouts of thermal systems in buildings, like smart thermostats. The beforementioned data collection and building operation model construction processes are most used in the following applications related to installed heating, ventilation, and air conditioning (HVAC) systems.

The first category of applications concerns fault detection in HVAC systems. For instance, ref. [20] recommends the identification of the gradual operation duration increase of cooling systems as an indication of the effectiveness and degradation of the heat exchanger connected to it. The second category of applications is related to HVAC operational optimization according to the user's needs. By using machine learning models, potential changes to the heating/cooling schedules are identified, aspiring to the maximization of the user's thermal comfort [21]. More specifically, it has been noticed that during instances when occupants enter the indoor space, the indoor air temperature may not be at the level desired by them. This causes thermal discomfort to them during their initial stay in the indoor environment. Similarly, there is an opportunity to save energy by prematurely turning off the heating/cooling systems before the indoor environment is going to become vacant [22]. These ways of utilizing thermal flexibility are achieved by applying optimization methodologies in the controlling aspect of energy management in buildings.

1.3. Limitations of Building Smart Controlling

Optimization cases require the training of a dynamic thermal behavior model of the indoor space. For instance, in the study of [23], a building (called 1-Resistance, 1-Capacitance model, 1R1C) and a system power model is generated using real operational data. However, such a complicated model structure may be too much for a smart thermostat to utilize. Not only that, but attempting to identify the characteristics of the building may be unnecessary for a controller, because a black-box model could do the same work just as accurately, while an interpretable model of the indoor space may not be useful to a user that has not the relevant knowledge background. Another characteristic application is the usage of model predictive controlling (MPC) for continuously adapting the model to its real counterpart (i.e., the building), as well as applying smart controlling strategies. However, this approach is usually accompanied by undesired computational cost that far surpasses the respective capabilities of a smart controller, since most models used resemble the deterministic model, such as that of [23]. Continuously updating the predictions means that simulations should be constantly performed. This means that predictive horizon-based approaches need a model that is easy and cheap to use in conventional programmable controllers.

A way to face this issue is to resort to training the building model on a cloud-based platform, which receives measurement data from the local smart thermostat sensors and sends back to it the recommended course of action that has been decided. This is performed in the work of [24], where a computer is used to simulate multi-zone models. It is also stated that equipping the MPC model into the controller is a challenge to face in their future work. In other words, the decision-making process and the computational requirements accompanied by it are transferred away from the local control, adopting a centralized structure, which requires a cost network connection to be sustained.

In order to decentralize building energy management, large computational requirements should be gotten rid of altogether. This can be achieved through a simple model structure that supports independent and localized training and decision-making capabilities in building energy systems. Moreover, the application of a simpler smart controlling structure, such as rule-based controlling, allows for a viable controlling solution while reducing the computational tax on the respective controllers [25]. For example, ref. [26] employs a fuzzy rule-based approach control, which features a controlling layout capable of managing the indoor air setpoint according to the optimal external environmental conditions while achieving simplicity and ease of implementation. Moreover, ref. [27] examined the prospects of combining a reinforcement learning (RL) control structure with a decision tree structure in order to increase its interpretability, thus forming an explainable Deep Q-Network structure. An interpretable RL agent is different than an interpretable building model, as the former describes which behavior encourages optimized objectives, which is desired information by every user. This led to a controlling agent that maintains its effectiveness towards energy savings while adopting a rule-based structure, which is more user-friendly and lightweight in its use than conventional RL structures.

The above research has uncovered two significant limitations regarding the operation of building smart controlling devices and, in particular, smart thermostats on the basis of optimal thermal load management in buildings. Firstly, most available controlling methodologies do not possess a simple and lightweight structure, so that they can be implemented by the thermostat itself, which causes concerns regarding its operational autonomy, reducing control from building users. Secondly, a simple yet accurate predictive model that is applied to those methodologies is equally vital to the independence and cost-effectiveness of a smart thermostat. While the model training task can be taken away from the smart thermostat, cases where the model is too complicated in structure hinder the predictive capabilities and consequently the decision-making speed of the thermostat. This leads to either the need for adopting a cloud network for this task or to increase the computational capabilities of the thermostat, both of which reduce its affordability in practical applications while also hindering controlling independency for each building.

Briefly speaking, smart controllers in building thermal load management applications are in need for simple models, as well as controlling structures and strategies. The first issue can be faced with the use of surrogate models [28], which are characterized by low computational requirements and satisfactory training and predictive performance of the thermal behavior of buildings, allowing low-cost smart thermostats to be developed. Their simple structure also allows model training sessions to be performed on site and re-used by the thermostat. Not only that, but their simplicity also allows for an effective combination of multiple surrogate models, each one trained for a specific behavior of the examined system, thus being able to predict system states jointly and potentially more accurately than a single-surrogate structure. These structures are referred to as multi-surrogate model in the literature [29]. Regarding the issue of controlling structure complexity, an answer to the high computational requirements of the MPC is given by [30], who propose that a machine learning model be combined with rule-based controlling strategies in a lightweight controlling layout. Their simulation results indicate that this is a step in the right direction, as both modeling through machine learning and the application of day-ahead planning through rule-based controlling (RBC) seem viable through this work. Still, concrete proof is needed for the fact that rule-based machine learning is suitable to handle building energy management tasks in an experimental approach. Moreover, the need to constantly train models that predict future building states in a localized manner can only be proved possible by displaying methods that can be equipped on programmable logic controllers (PLCs) in an experimental environment as well. A future smart thermostat that not only supports offline training possibilities, but also online ones, will offer autonomy and security of operation of the building energy management system it is equipped on.

Besides the limitations of the scope due to the nature of the models required by modern smart controllers and thermostats, the current research is also limited in experimental applications of smart thermostats implemented in real-time operation cases. On the other hand, an ample number of works have explored thermostat data and developed guidelines on the effect of scheduling and controller programming on energy savings and thermal comfort levels [17,31,32]. Moreover, thermostat data have also been utilized in simulation environments, either to generate building models suitable for day-ahead predictions in MPC applications [33], to create occupancy model generators [34], or to directly estimate the characteristics of the building through time constant evaluation [33,35].

Two works worth mentioning that utilize a first-order building model to simulate a single-room indoor space are those of Sun et al. [36,37], which estimate the preheating time required to set a building room to desired temperature levels at the start of the occupancy. The preheating time estimation in these works is accompanied by an uncertainty period. Two limitations arise from this set of works. The first one is that the preheating time is estimated, and not decided by a controller, as the methodology itself contains a preheating period probability estimation. In order for the preheating algorithm to be implemented practically, this uncertainty needs to be gotten rid of, since a controller needs to make a concrete decision. Secondly, those works implement simplified first-order white-box models, which are too computationally costly for a programmable controller to identify the model parameters based on operational building data, as established in the previous sections. In short, those limitations that are found among the literature need to be addressed.

1.4. Aim of This Work

This approach aspires to a methodology for developing low-cost smart thermostats that maintain autonomous and effective optimized controlling capabilities of the heating schedule, without the need for either cloud based controlling or computational requirements that exceed those of a conventional PLC. Moreover, the presented smart thermostat methodology can be extended to include supply-side management, which consists of the direct control of the heating system itself, providing prospects for a demand-and-supply load management methodology, aiming for the optimized utilization of RES systems, the efficient use of HVAC systems, and the achievement of desired thermal comfort levels at the same time. In other words, the proposed approach attempts to face the current limitations of the existing smart controller development methodologies that have been spotted in the literature, which relate to the restriction of the thermostat training and predictive capabilities to be provided remotely, consequently binding a smart thermostat to a centralized database. Oppositely, enabling the thermostat to predict the building state by itself and helping decision support adopt a more autonomous controlling layout is something desirable in current standalone heating applications. This also holds true for future district heating network (DHN) applications, where thermal loads will be able to be provided in a more distributed manner, surpassing the limitations of centralized thermal load management layouts currently implemented on an urban scale [38]. Developing and testing smart thermostats in practical experimental applications allows for matching the nature of the controlling hierarchy of future DHNs to their distributed thermal load production and provision. This is because developing such devices with autonomy will allow various benefits of distributed energy system networks to be implemented to DHNs, such as the inclusion of prosumers in the network, enhanced renewable-energy-based thermal load self-consumption, and security of supply. Moreover, there is an increased sense of control for the DHN users and more direct communication with the respective producers, leading to the evolution of demand-and-supply-side management techniques. Considering that urban-scale energy networks and energy management strategies are promising structures to be used even more in the near future, the development of autonomous controlling applications through experimental applications such as this work is important.

In a brief description of the presented work, which faces the above issues, it focuses on developing and validating a smart thermostat layout that features a lightweight model and a demand-side management methodology in the form of a rule-based preheating controlling strategy. Methods for generating simple yet accurate models with low computational requirements, such as that of ALAMO modeling [39], paired with a lightweight controlling strategy approach, such as the RBC approach, is a great combination to address not only the reliance of smart thermostats on cloud computing but also to get rid of the large computational loads altogether. Not only that, but the automated machine learning methods (also stated by [30]) such as the ALAMO approach surpasses the hyperparameterization needs posed in other machine learning methods, as well as simplifies both the installation and operation of machine learning libraries into programmable controllers. In short, the two significant limitations posed in the works [36,37], as well as the limitation of scarcity of methodologies that are developed and applied experimentally in real-time operation are faced in this work.

Moving to a more specific analysis of the proposed work, the embedded surrogate model is trained on the thermal behavior data of the environmental chamber, while the smart thermostat employs it in real time to predict the indoor conditions of the environmental chamber and apply the controlling strategy according to the optimization criteria set by a building occupant. Those are the maximization of thermal comfort of the occupant and the use heating systems of the chamber as little as possible. The following sections describe the materials used for the development of the proposed methodology, the results that were produced through its experimental application, as well as the future potential of the approach in thermal networks and the conclusions drawn.

2. Materials and Methods

2.1. General Description

In this section, the methodology of smart thermostat development and validation process is described. The general idea of the smart thermostat is to create a controller that is able to predict the future conditions in the environmental chamber and take necessary actions regarding the heating schedule in advance. Following the order of actions presented in the flow chart of Figure 1, the methodology initiates by collecting measurements linked to the operation of the experimental chamber. These measurements are the indoor air temperature of the chamber and the heating operation signal that correspond to the operation of the chamber during the time period 22 February 2023–28 February 2023. Since the experimental chamber is protected from outside weather conditions, the most prevalent parameters that affect the indoor temperature inside the chamber were deemed to be the previous indoor temperature, as well as the use of heating. Therefore, those two parameters will probably be the most suitable ones for the creation of the surrogate model. The creation of such a data batch enables its use on the training and validation of a surrogate model using a lightweight machine learning software called ALAMO (version 2024.10.25). Data collection was conducted by the controller itself, as it has embedded data-logging capabilities in its default software. The measured data are automatically filtered by mean value filters to mitigate measurement noise. The quality of data were sufficient for this case, so no further preprocessing took place at this stage. The rest of the requirements for providing predictive capabilities to the thermostat is for it to possess information regarding the chamber occupancy schedule, as well as the desired thermal comfort settings of the occupant. Namely, those are the indoor air temperature setpoint and the daily occupancy schedule. The ability of the model to predict is used to optimize the operation of the heating schedule.

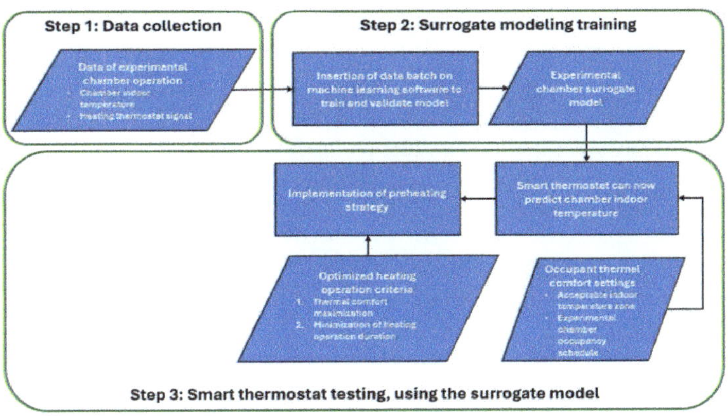

Figure 1. Flow chart depicting the smart thermostat development and validation process.

The primary objective of the smart thermostat in this case is to maximize the thermal comfort of the occupant, while the secondary objective is to minimize the necessary operation duration for this task. For this purpose, a preheating controlling strategy is employed, which will be analyzed in Section 2.2.4. It is noted that in this experimental case, the energy consumption measurements are not available and therefore, the operation duration of the heating system is planned by the thermostat to be as little as possible.

In order to verify the effectiveness of this methodology, an operation validation phase takes place, which includes the following steps:

1. Operational data gathering (22 February 2023–1 March 2023);
    a. Occupancy/heating schedule: 21:00–01:00;
2. Surrogate model training phase using gathered operational data;
3. Real-time smart thermostat testing, equipped with the trained surrogate model, during the period 14 February 2024–16 February 2024. The characteristics of the operation during test operation are the following:
    b. Occupancy schedule: 18:00–21:00;

c. Heating schedule: 18:00–21:00 + preheating period (to be calculated autonomously in real-time by the thermostat);
d. Environmental chamber temperature setpoint: [19.5–20.5] °C.

2.2. Experimental Chamber

2.2.1. Chamber Characteristics

In this section the environmental chamber is presented, of which the operational data is collected (Figure 2). This chamber can be characterized as a box-in-a-box structure and is situated in the facilities of University of Western Macedonia campus near Koila, Kozani, Greece. Table 1 displays the dimensions of the experimental chamber, which are close to that of typical building rooms that are used in existing works in the present literature. The box-in-a-box structure allows for achieving a controlled environment inside the chamber, while the air conditioners outside the chamber have slow transitions with small fluctuations, mitigating the effect of outdoor conditions. This also means that the effects of direct sunlight on the chamber are mitigated, as it is entirely covered by the surrounding building environment. In other words, a potential simulation of the environmental chamber is easier than the case of real buildings, as many parameters and thermal mechanics are not considered to be fluctuating during the simulation.

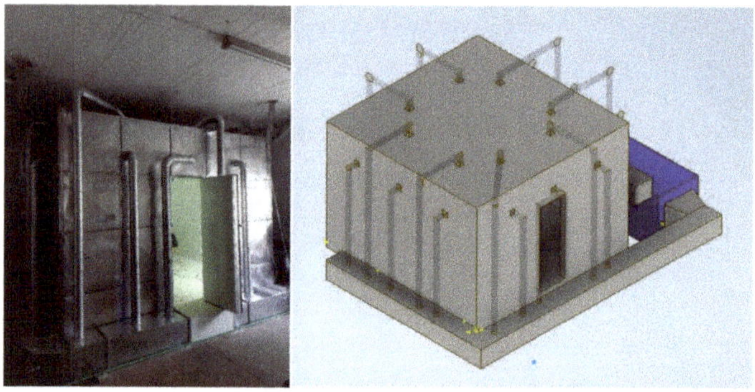

Figure 2. Environmental chamber.

Table 1. Environmental chamber general characteristics.

Dimensions		Units
Height	2.85	m
Length	4.02	m
Width	4.02	m
Wall thickness	0.135	m
Mean heat loss factor	0.423	$W/(m^2{\cdot}K)$
Total thermal capacity	$6.46{\cdot}10^5$	J/K

The control of the indoor chamber environment is achieved by a heating system (Table 2). More specifically, it consists of an air duct system, where a heat pump is also embedded. The latter is responsible for providing heating or cooling loads to the chamber, depending on the thermal demands of the indoor chamber environment. The nominal heating power capacity for this experiment is stable at 4.7 kW, since on/off heat control is implemented. Besides heating and cooling, the use of an air duct system enables the import of fresh air in the environmental chamber, as well as configuration of the relative humidity of the indoor air through water spraying. It is noted that the air duct provides an air supply that reaches 1725 m^3/h. While the import of fresh air is possible, no fresh air is imported in the current experimental case. Moreover, water spraying allows for controlled humidification that can lead to relative humidity levels reaching 90%, although this study case does not include air humidity control and therefore, the relative humidity levels remain mostly stable around 40% throughout the experimental periods mentioned in Section 2.1.

Table 2. General characteristics of the chamber heating/humidity control system.

Characteristic	Value	Unit of Measurement
Heat pump thermal capacity	4.7	kW
Water spraying relative humidity range of control	40–90	%
Air duct air supply	1725	m^3/h

2.2.2. Smart Thermostat Setup

The smart thermostat setup consists of a PLC layout, which is programmable using embedded Python (MicroPython version 1.14) environment. The controller layout consists of a "master" controller which is linked to "slave" micro-controllers, which are responsible for providing the signal inputs and output that the "master" controller instructs, as well as the collection of measured data with data filtering equipped to prevent excessive measurement noise. Figure 3 displays the controller that is operating as a smart thermostat during this experiment. Additionally, Figure 4 displays the setup configuration of the controllers used, along with the input and output signals that are utilized from the controller, when acting as a smart thermostat. Table 3 displays the specifications of the programmable controller. As can be seen, the controller has pretty low computational power with the additional benefit that it enables the user to freely implement various controlling strategies through coding in the embedded Python environment. Moreover, it possesses inherent data logging capabilities and the ability to be equipped with external Python libraries, such as available open-source ALAMO approach libraries in Python environment. The above characteristics not only allow the controller to host externally trained surrogate models and use them in predictive controlling strategies, as it will be made clear in Section 3, but is also expected to be able to host the ALAMO surrogate model training process in future efforts, should it be equipped with the respective ALAMO Python (ALAMOPY) library.

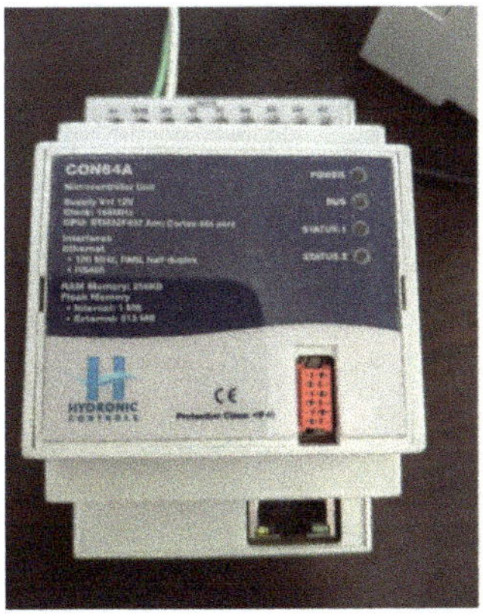

Figure 3. Programmable controller operating as smart thermostat.

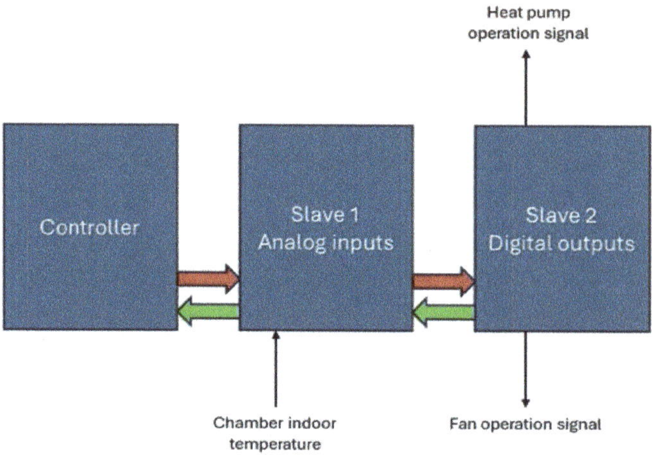

Figure 4. Controller setup in the current work. Arrows indicate the directions of data and signals.

Table 3. Programmable controller specifications.

Specification	Value
CPU model	STM32F437 Arm Cortex-M4 core
Clock Speed	168 MHz
RAM memory	256 KB
Internal flash memory	1 MB
External flash memory	64 MB

During the experiment, the controller is equipped with the surrogate model that is created using collected temperature and heating signal data, so that it can essentially become capable of predicting the indoor air temperature of the environmental chamber and attempting to maximize thermal comfort, by adding a preheating period, besides its conventional heating operation.

2.2.3. Chamber Surrogate Model

Surrogate models are able to imitate the operational behavior of the system they are trained on. In the experimental case, the surrogate models that are created attempt to depict the fluctuation of temperature of the experimental chamber throughout its operation. In order to create those models, the software called ALAMO (Figure 5), which is based on the ALAMO approach [39], is used.

In this experimental case, two surrogate models are going to be created and compared, in order to use the most accurate one. More specifically, the surrogate models that are created have structures that are displayed in Equations (1)–(3). Both models are trained using the indoor air temperatures (T_{in}) and the heating activation signal sent by the controller-thermostat (U_t). The general idea of the models is to estimate the chamber indoor air temperature during the next simulation timestep $T_{in,t+dt}$, by using the temperature in the current timestep ($T_{in,t}$), as well as the chamber heating request signal (U_t). The first surrogate model "f_{mono}" is a single-surrogate model, as it possesses a single equation structure and is trained on the whole temperature data set.

$$T_{in,t+dt} = f_{mono}(T_{in,t}, U_t) \quad (1)$$

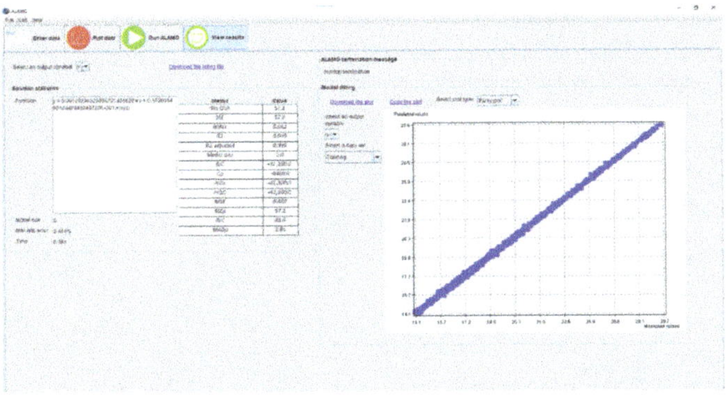

Figure 5. ALAMO environment.

On the contrary, the second model is a multi-surrogate model. The logic behind the multi-surrogate model is that multiple models are trained on a part of the dataset, which satisfies certain conditions. Therefore, during prediction, the multi-surrogate model employs a different sub-model, depending on which of the conditions hold true during the predictive horizon. This allows for having more specialized sub-models and, therefore, improving predictive accuracy.

To specify, two surrogate model equations are used in this model, with each one training on a part of the dataset. As Equations (2) and (3) display, one model (Equation (2)) is trained on the part of the dataset, in which the heating request signal is absent, resulting in absence of heating in the chamber. Essentially, the dataset used to train this model captures the free-floating behavior of the temperature in the chamber. On the other hand, Equation (3) is trained on part of the dataset, where heating of the environmental chamber is activated.

$$T_{in,t+dt} = f_{s1}(T_{in,t}, U_t),\ U_t = 0 \tag{2}$$

$$T_{in,t+dt} = f_{s2}(T_{in,t}, U_t),\ U_t = 1 \tag{3}$$

The fitting criterion in all the surrogate model training cases is the Root Mean Square Error ($RMSE$) (Equation (4)) and the Mean Absolute Error (MAE) (Equation (5)) between the measured and the predicted indoor air temperature value of the experimental chamber.

$$RMSE = \sqrt{\frac{\sum_{i=1}^{N} T_{in,p,t} - T_{in,m,t}}{N}} \tag{4}$$

$$MAE = \frac{\sum_{i=1}^{N} \left| T_{in,p,t} - T_{in,m,t} \right|}{N} \tag{5}$$

2.2.4. Proposed Smart Thermostat Strategy

The main characteristics that set the proposed smart thermostat apart from programmable ones are the following:

- Heating periods of the indoor space are independent from the building occupancy periods and are decided by the smart thermostat.
- Since the smart thermostat is free to act independently, regarding the heating period configuration, indoor air temperature does not concern the user anymore. In other words, $T_{setback}$ is no longer a configurable parameter, whereas $T_{setpoint}$ can still be used.
- Heating periods are now configured by the smart thermostat according to the value of $T_{setpoint}$ the user has provided.

- The proposed smart thermostat is able to predict the indoor air temperature using model training methods and data (i.e., indoor air temperature, heating request signal).

The smart thermostat heating strategy, which is used daily, is presented below (Figure 6). Essentially, this algorithm uses the moving horizon predictive control, which is commonly used among model predictive control (MPC). However, in this case the length of the moving predictive horizon is not constant, as in most MPC practices, but is variable and is calculated according to the temporal distance ($\Delta\tau$) between the present moment and the moment when occupancy starts, which also coincides with the start of heating (Equation (6)).

$$\Delta\tau = t_{turn\_on} - t_{now} \tag{6}$$

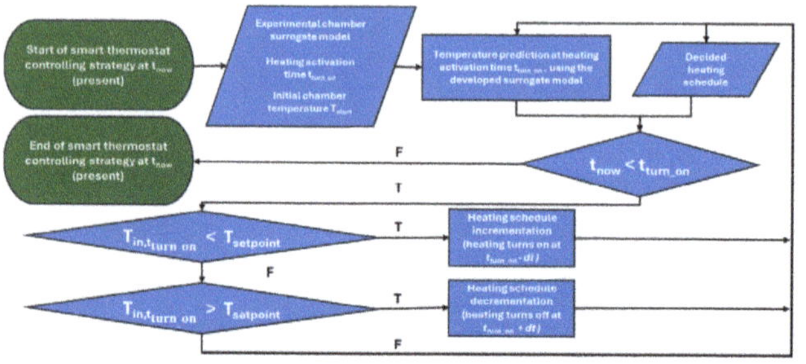

Figure 6. Smart thermostat heating algorithm used to heat the indoor air of the chamber. Letters "T" and "F" stand for "True" and "False", respectively.

Note that this is an initiation of the heating turn-on scheduling and is subject to changes when preheating algorithm is used. In other words, when the smart thermostat algorithm initiates, the heating schedule is the same as the occupancy schedule. Then, the preheating strategy is applied, and an additional preheating period is added. The variable moving horizon strategy ensures a satisfactory initial estimation of the preheating schedule, which becomes more accurate as the moving horizon length is reduced, namely as the present moment t_{now} approaches the heating enabling moment t_{turn_on}. This implementation is necessary for time-dependent problems like temperature prediction, as the indoor temperature each moment (else called timestep in simulations) depends on the value of the moment before that. Therefore, as the thermostat predicts further into the future its prediction accuracy reduces, and vice versa. This approach makes the prediction robust and independent of the initial indoor temperature conditions after the occupancy period ends and preheating duration calculation initiates.

Regarding the preheating strategy of the smart thermostat, the procedure is initiated by defining the occupancy schedule of the user. Consequently, the time of occupant's arrival ($t_{arrival}$) is set as the starting point of conventional heating schedule. From this point on, the thermostat uses the developed surrogate model and the indoor air temperature at present to predict the temperature for every time interval until time interval $t_{arrival}$. Depending on whether the predicted temperature reaches the desired temperature setpoint, the thermostat adjusts the heating schedule by configuring the preheating schedule in a continuous manner.

This algorithm has been formed with the aim of maximizing the occupant's thermal comfort, while also minimizing the heating schedule size. It is also noted that the current approach focuses entirely on optimally managing the demanded heating load, while ignoring any potential changes in thermal energy production efficiency or cost that could occur depending on the chosen schedule. In other words, this methodology does not consider the supply-side management of the heating energy used for the environmental chamber.

The result is a formed heating schedule which resembles that of Figure 7. As observed in the figure, the heating schedule is divided into two areas: the occupancy period (green area), and the preheating period (red area). The proposed smart thermostat aims at maintaining the indoor air temperature in the thermal comfort zone (yellow area), and utilizes the preheating controlling strategy to do so. As it is also evident in the figure, the thermostat still operates as a conventional one during the occupancy period, turning the heating off when the indoor air tends to be overheated or overcool, i.e., turning on and off the heating as the temperature approaches the lower and upper thermal comfort temperature limits, respectively.

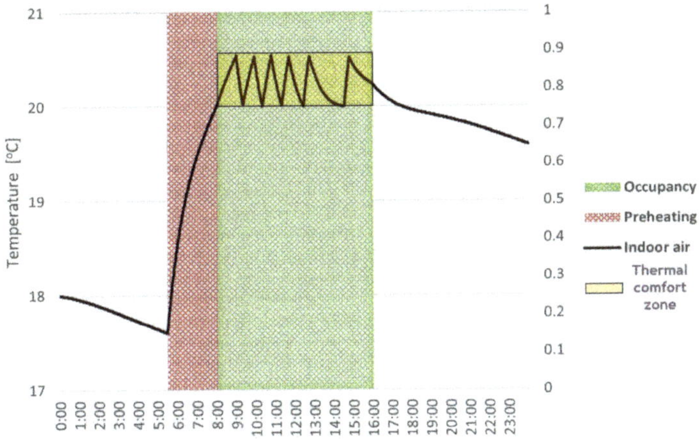

Figure 7. Example of implementation of the smart thermostat controlling algorithm.

Preheating Strategy

As already mentioned, the use of a preheating strategy is crucial to the maximization of thermal comfort and has been proven in our previous work that it is able to maximize thermal comfort levels to 100% [18], as well as increase the thermal flexibility of buildings. This strategy has also been used by [36,37] to increase the thermal comfort and counter the presence of thermal inertia that leads to thermal discomfort during the first moments of occupancy. Figure 8 displays the algorithm of the preheating strategy used by the smart thermostat.

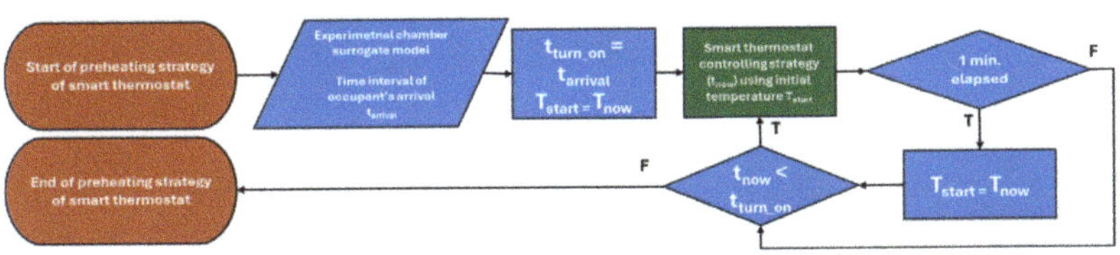

Figure 8. Smart preheating strategy algorithm. Letters "T" and "F" stand for words "True" and "False", respectively.

The preheating algorithm initiates when the controlled indoor space becomes unoccupied. Then, the next occupancy period is examined, and the respective preheating period begins to form. This is done by initiating the heating period to be the same as the occupancy period. Therefore, as displayed in Figure 8, the moment that the occupant enters the controlled indoor space, $t_{arrival}$, is equal to the time that the heating period initiates.

Then, the smart thermostat uses the surrogate model to predict the indoor air temperature using the current heating schedule, depending on whether the indoor air temperature when the occupant arrives during $t_{arrival}$ deviates from the lower bound of the occupant thermal comfort zone temperature, which is also considered the chosen temperature setpoint in this case. It is reminded that the current approach focuses on the demand-side load management, namely the heating loads requested by the energy consumer, ignoring potential energy consumption cost, as this is included in the supply-side management.

Implementing the Smart Preheating Strategy

For the purpose of validating the methodology that is used by the proposed smart thermostat, the most representative surrogate model (i.e., the one with the lowest *RMSE* during the training phase) is embedded in the thermostat and smart preheating controlling strategy is utilized for three consecutive days. The heating optimization aims to maximize the thermal comfort, while requesting as little heating as possible. Therefore, daily occupant thermal comfort TC is calculated according to Equation (7). More specifically, thermal comfort is calculated as the percentage of the n minutes contained in the occupancy period, during which indoor air temperature $T_{in,i}$ (for each timestep i) is bigger than the temperature setpoint $T_{in,set}$ that the user has chosen. It is noted that while comfort zones have lower and upper temperature bounds, potential overheating may be observed during occupancy periods, which skew the evaluation of the smart thermostat performance. This is because temperature overheating is the result of heating turnoff delay which may be linked to sensor hysteresis. In any case, thermal comfort is evaluated in two ways, where the Equation (8) considers indoor air temperature overheating due to sensor hysteresis, while Equation (7) does not. The overheating effect will be observable in the smart thermostat testing phase of the Section 3.

$$TC = \frac{\sum_{i=1}^{N}(T_{in,i} \geq T_{in,set})}{n} \tag{7}$$

$$TC_o = \frac{\sum_{i=1}^{N}[(T_{in,i} \geq T_{in,set})\ AND\ (T_{in,i} \leq T_{in,set}+1)]}{n} \tag{8}$$

In addition, since no measurement regarding energy consumption is available during this experiment, the minimization of heating operation is described in operational hours.

3. Results

3.1. Operational Data Gathering

In this section, the implementation of the experimental methodology is presented. To begin with, the measured data that are used in the training of the surrogate models are related to the operational period of the environmental chamber 22 February 2023–1 March 2023. Figure 9 presents the measured data, namely the chamber indoor air temperature and the heating operation signal.

As can be seen from Figure 9, heating periods take place during 21:00–01:00 each day, while no temperature setpoint is chosen, in order to achieve uninterrupted heating of the chamber. Consequently, the indoor air of the heated environmental chamber reaches 30 °C, and then after 01:00 it enters the free-floating temperature state, where it gradually cools down, due to heat thermal losses. An interesting observation is that the lowest temperature the chamber reaches each day gradually increases after day, starting at 13.5 °C and finishing at 17.5 °C at the end of the sequence of days. This might be caused by either of two reasons: (a) The environmental chamber walls gradually build up heating energy each day, leading to the indoor air not dropping to the same temperature that it was before heating, or (b) the hysteresis of the measuring instrument is affected by the consecutive heating and cooling of the air during the examined period, leading to measurement errors [40].

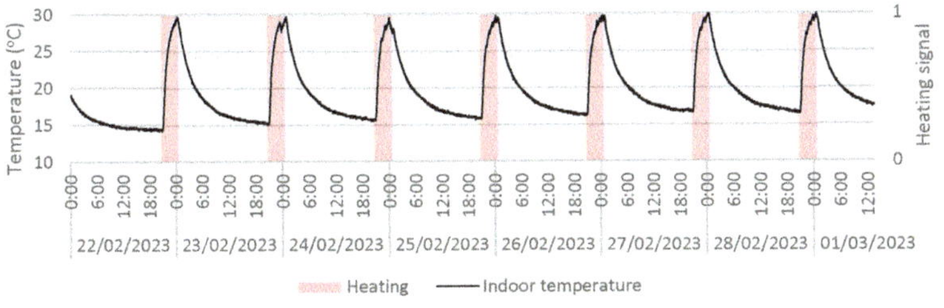

Figure 9. Collected measurement data.

3.2. Surrogate Model Training

Since data are collected, the training of the proposed surrogate models follows. By using the ALAMO software and examining the behavior of the temperature in the experimental data batch, the model structures are created, and coefficients are calculated by data fitting. Equations (9)–(11) below are provided, where Equation (9) represents the single-surrogate model (f_{mono}), whereas Equations (10) and (11) represent the multi-surrogate model (f_{s1}, f_{s2}). As elaborated in Section 2, both models describe the temperature during the next timestep and depend on the indoor temperature during the previous timestep, as well as whether the heating signal is on (f_{s1}) or off (f_{s2}).

$$f_{mono} : T_{in,t+dt} = 0.140 * U_t + 0.959 * T_{in,t} + 0.693 * ln(T_{in,t}) - 1.266 \quad (9)$$

$$f_{s1} : T_{in,t+dt} = 0.920 * T_{in,t} + 1.458 * ln(T_{in,t}) - 2.606 \quad (10)$$

$$f_{s2} : T_{in,t+dt} = 0.993 * T_{in,t} + 0.038 * \ln(T_{in,t}) \quad (11)$$

The metric values for each model training phase are also provided in Table 4 below. As observed, the produced models have satisfying accuracy. However, the multi-surrogate model containing specialized sub-models gives it an extra edge in performance. While the differences are almost indistinguishable, propagating errors throughout time horizon predictions is able to increase the magnitude of errors among models, as it becomes clear below, in Figure 10.

Table 4. Data training results of the surrogate models.

Model	*RMSE*	R^2	*MAE*
f_{mono}	0.096	1	0.028
f_{s1}	0.116	0.999	0.021
f_{s2}	0.082	0.999	0.028

In order to understand the difference in performances of the surrogate models, a demonstration of them predicting the chamber operation is presented in Figure 10. More specifically, each of the models has a predictive horizon of 24 h, which means they predict the whole day of operation by having available only the initial indoor air temperature of the chamber of each day. As can be seen, the multi-surrogate model (f_{s1}, f_{s2}) performs satisfactorily. On the other hand, the single-surrogate model is able to capture the behavior of the indoor air temperature during heating periods, the same cannot be said for the free-floating periods. It should be noted that since the surrogate model predicts the indoor air temperature accurately during the next time interval, which is 1 min, high accuracy is expected only when predicting the next few hours. This means that *RMSE* metrics produced by the model fitting procedure that corresponds to the training performance may not provide a clear picture of predictive performance in cases of extended prediction horizon. This is why comparing models on the training dataset with large predictive horizons, like in Figure 10, gives a better image of which model is more suitable for application. In any case, as explained in Section 2, the predictive horizon used is not going to be static, so the predictive performance of the surrogate model used is going to improve as the predictive horizon reduces in size. This means that such accuracy in a 24 h horizon (Figure 10) is more than enough for the smart thermostat to form the preheating period effectively during the smart preheating implementation phase.

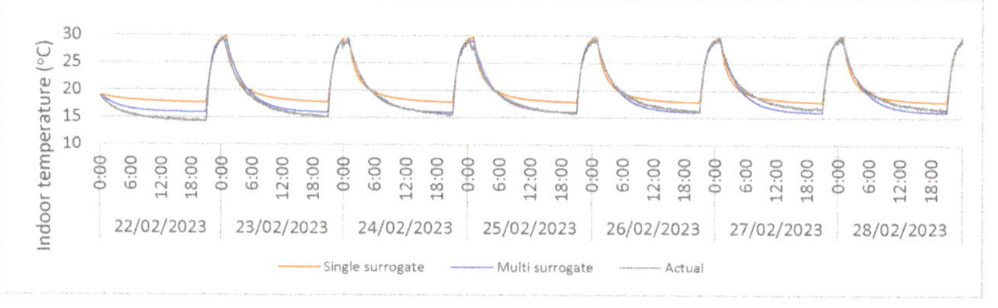

Figure 10. Visual comparison of the surrogate model performance.

Initially, it may seem contradictory that models with similar *RMSE* metrics may have different prediction performances when the predictive horizon is large. To expand on this, Figure 11 displays the *RMSE* metrics when the prediction horizon length $\Delta\tau$ gradually reduces in size, for both the single-surrogate and multi-surrogate models. Essentially, Figure 10 displays the temperature predictions of both models when $\Delta\tau$ is equal to 1440 min (24 h), whereas when $\Delta\tau$ equal to 1 min, the *RMSE* values are the similar to those in Table 4.

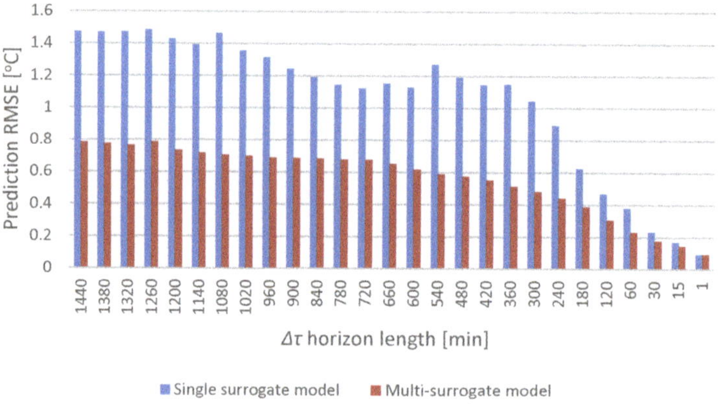

Figure 11. Sensitivity analysis of predictive horizon length on the predictive performance of both models when using the training dataset.

By examining Figure 11, it can be deduced that the multi-surrogate model tends to predict better, while it has a more robust performance than the single-surrogate model, as horizon length decreases. This might be attributed to the fact that the multi-surrogate model contains sub-models that are specialized in predicting different building behavior types (i.e., heating and free-fall periods), while the single surrogate is less specialized in each behavior, leading to it predicting less accurately. Moreover, it can be illustrated that as the horizon is reduced in size, the prediction propagation error also decreases for both cases. This means that even a slightly less accurate prediction will improve as t_{now} approaches $t_{arrival}$, which has also been mentioned in Section 2.2.4. This is why the variable predictive horizon provides robustness to the preheating algorithm, regardless of the initial temperature and length of the predictive horizon. Furthermore, this also explains the predictive performance difference of the models in large predictive horizons, even if the training metrics are the same. Regardless, while the surrogate model was chosen as more suitable in this case, the authors cannot confidently state that multi-surrogate models are better in every case, as this requires a more systematic model comparison, which is not the focus of the current work. Consequently, the multi-surrogate model will be used by the smart thermostat during the testing phase.

3.3. Smart Thermostat Testing

After finishing the surrogate model training and evaluation process, the testing phase of the smart thermostat takes place. As mentioned in Section 2, the occupancy schedule period is 18:00–21:00, while the preheating period is calculated by the smart thermostat in real time using the temperature prediction and smart preheating strategy. It is noted that while the occupancy schedule of the testing phase differs from the respective schedule of the training period, this poses a good opportunity to test the generality of the developed model, namely it being able to predict and calculate the preheating periods at any given time of day. The resulting test is displayed in Figure 12, where the thermostatic control is applied in real time. More specifically, the conventional heating periods (Heating) are taking place in the occupancy period, while preheating periods manifest before the occupancy periods, as intended. Three consecutive days of operation display three characteristic operational versions of preheating through prediction.

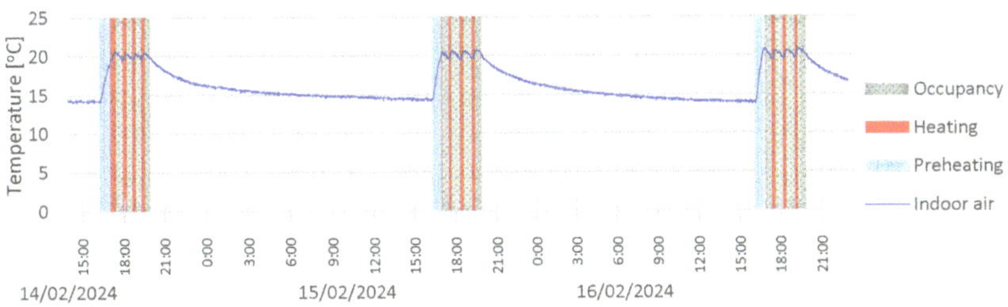

Figure 12. Testing smart thermostat operation in real time.

Starting with 14 February (Figure 13), a slight delay of the preheating start is manifested, which can be spotted because of the heating that takes place during the start of the occupancy phase at 18:00. This means by the time the occupant would have entered the indoor space, the indoor temperature had not reached the comfort desired value according to the comfort zone [19.5–20.5] °C. Therefore, thermal comfort of the occupant was calculated to be 97.2% for this occupancy period, when overheating is not considered. In this case, the overheating was almost negligible, as indoor air temperature surpassed the upper temperature bound during the occupancy period for a very short interval just before 18:30.

On the other hand, the exact opposite phenomenon occurred on 16 February (Figure 14), where preheating mode was activated slightly earlier than the optimal, resulting in the indoor temperature reaching comfortable temperature conditions, though this leads to the chamber indoor air being maintained at higher temperature levels at the start of the occupancy, leading to increased heat losses and therefore, more energy required to maintain comfortable temperature levels. On another note, slight overheating is spotted, which may be attributed to the temperature sensor hysteresis. This overheating essentially manifests when the indoor air temperature (black line) surpasses the upper temperature bound (green line) during the occupancy period.

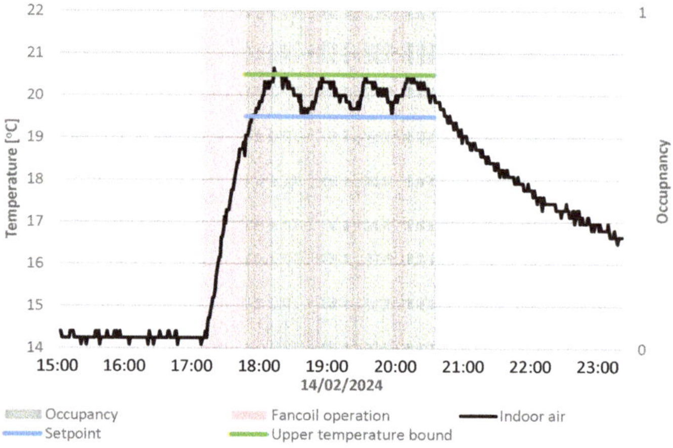

Figure 13. Testing phase, 14 February 2024.

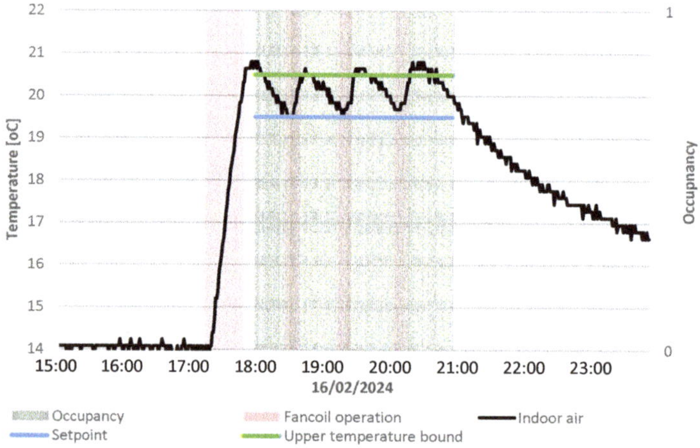

Figure 14. Testing phase, 16 February 2024.

Finally, on 15 February (Figure 15), preheating leads perfectly to the indoor temperature being at the lower end of the occupant thermal comfort zone (19.5 °C) at the start of the occupancy period. In addition, slight overheating of the chamber is also spotted in this case, when indoor air temperature (black line) surpasses the upper temperature bound (green line), during the occupancy period. What is also interesting, however, by carefully examining 15 February (Figure 15) and 16 February (Figure 14), it might be noticed that the operation periods of the heating may not differ so much, in contrast to 14 February (Figure 13). This is verified by examining Table 5, where the occupancy and the heating periods (fan coil operation) are displayed.

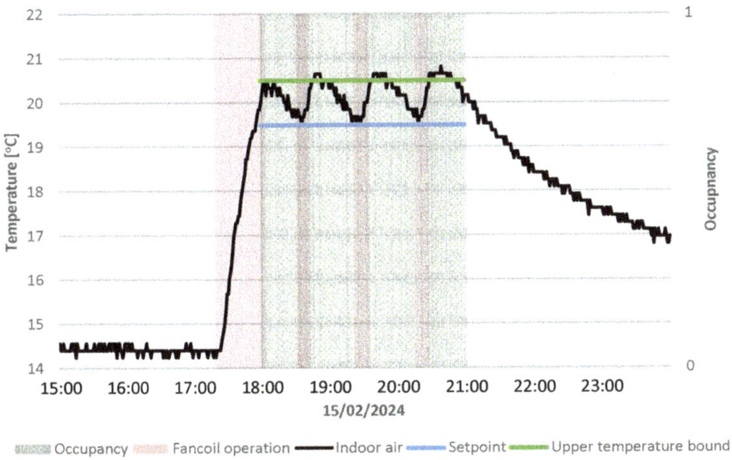

Figure 15. Testing phase, 15 February 2024.

Table 5. Characteristic periods and results of chamber operation during the testing phase.

Date	No. of Occupancy Minutes	Fan Coil Operation (Heating) Minutes	Thermally Comfortable Minutes (Hysteresis Considered)	Thermally Comfortable Minutes (Hysteresis Not Considered)	Thermal Comfort (*TC*) (Overheating Considered) [%]	Thermal Comfort (*TC*) (Overheating Not Considered) [%]
14 February 2024	179.00	110.00	173.00	174.00	96.65	97.20
15 February 2024	178.00	82.00	149.00	178.00	83.71	100.00
16 February 2024	178.00	75.00	145.00	178.00	81.46	100.00
Standard deviation in period 14 February 2024–16 February 2024	0.47	15.12	12.36	1.88	6.69	1.32

By examining Table 5, it can be seen that 14 February involves larger heating operation periods, as temperature increases more slowly in the chamber. If the indoor temperature had risen at the same rate as in the other days, then there would have not been such a great difference from the other testing days. At this time, the cause of such a difference is not known, and it might be attributed to faulty sensor readings. On the other hand, 15 February and 16 February display consistency in their results. Despite that, 16 February has reduced heating operation, as indoor temperature during occupancy period increases at a faster rate than the case of 15 February. Therefore, even in a controlled environment, temperature sensor characteristics play a crucial role in successfully implementing heating strategies. It should be noted that this fan coil operation reduction across 14 February–16 February indicates a faster temperature increase of the indoor air. While this is a phenomenon worth looking into, the main scope of this work is the ability of the smart thermostat algorithm to maximize thermal comfort levels; therefore, this is something that should be investigated in the future.

4. Discussion

Generally speaking, the downscaling of the computational demand required for model training, as well as decision-making through the use of controlling strategies, is also a great benefit in multiple aspects. First and foremost, the economic aspect: Namely, the investment cost has the potential to drop significantly, making smart thermostats a much more affordable tool used in either residential or service buildings, where most people spend their time nowadays and, subsequently, most building thermal load requests manifest [1]. Secondly, the recent focus of the EU on developing large-scale RES-based communities [7] to secure energy provision to buildings has opened discussion on the layouts that support its successful implementation, such as DHNs. It has been discussed in [38] that such layouts are possible, while it is also hinted that heating distribution in latest-generation DHNs will possibly adopt a decentralized layout, enabling autonomy between heating producers. This leads to an opportunity to help the DHNs to include a multi-energy thermal market, where each producer and each consumer has its own autonomy and independent aims in terms of decision-making. This transition from a centralized DHN to distributed DHNs will also benefit the inclusion of RESs, as not only will RES producers have individualized aims and courses of action, but distributed systems will also allow anyone to become an energy producer at any time, encouraging "prosumers", namely consumers that can also produce energy (usually through locally installed RES systems) to participate in the thermal energy market of the respective DHN. Nevertheless, for all that to be possible, building consumers should also have high levels of decision-making autonomy. This means that a controlling structure in such systems should also be distributed equally among producers and consumers. Therefore, the development methodologies of independent smart controllers, such as the one in the proposed work, strive to be both affordable and implementable in a distributed controlling layout, bringing substantial contribution value to the table.

According to the observed results, it can be inferred that developing simplified smart controlling tools in practical applications is implementable. When comparing the works mostly related to the proposed one, namely that of [36,37], various challenges have been faced. To begin with, this work implements a model-equipped controller that automatically predicts the future states of the indoor air temperature, which was not conducted in those works. The use of surrogate modeling surpasses the limitations of the physical models used in these works because it gives the controller the potential to create simple models by itself, as well as use them in real time. Furthermore, the use of simple models allowed us to face the second limitations of the referred works: They can be used to continuously predict and adjust the future heating scheduling by utilizing a variable predicting horizon. This is something that has been utilized in MPCs and proves the potential of that method when paired with models that are easy to simulate with. This ability to continuously predict the future state of the indoor air allows for updating the preheating schedule as the time to enable heating approaches. Therefore, there is no need to evaluate uncertainty, as the predictive uncertainty mitigates when the preheating period comes closer to the present and the predictive horizon reduces in size, making the prediction more accurate as time passes by. In short, the controller initially estimates the heating period, but in contrast to [36,37], the estimation slowly transitions into a decision to be taken by the controller. All the above improvements allowed the preheating strategy to be evolved to be suitable for use in real-time applications. Finally, it should be noted that while the environmental chamber is not affected by the outdoor environment, its indoor space is similar in size to the indoor spaces that the methodologies of [36,37] are tested on.

Regarding other benefits this work brings, setting a basis for model development on an environmental chamber such as that of this work helps bring forth potential modeling issues (such as dealing with sensor drifts, which are identifiable in this case), which are difficult to observe in real building cases, where the air temperature behavior is affected by more parameters. For example, even in a controlled environment such as the current environmental chamber, an increase in heating duration was displayed on 14 February, which may be attributed to possible energy accumulation inside the chamber envelope.

Performing experiments on days that are not consecutive, as well as including wall temperature measurements of the environmental chamber, could potentially answer whether this effect is linked to heat accumulation in the chamber envelope. This is something that should be done before applying the proposed smart thermostat methodology to more complex buildings. This helps ensure the reliability of the developed models in future efforts, as well as provides insight into issues likely to manifest during the validation of a smart thermostat, which are more difficult to discern in a real building operation. On another note, despite the existing measurement drift spotted in the training phase, the model prediction and the preheating were mostly successful, indicating it did not affect the implementation of the methodology that much. Regardless, whether this drift results from either sensor hysteresis or the thermal capacity of the environmental chamber needs to be identified in the future. Future energy consumption measurements will be able to give a more concrete indication of which of the two is true. Furthermore, those measurements allow for estimating the potential energy savings of the implemented controlling strategies. Therefore, this is considered one of the high-priority future tasks of this work. Additionally, the presence of sensor hysteresis when consecutively heating during occupancy periods, which leads to occasional overheating, is not connected to the performance of the preheating algorithm in this case. However, it is something that should be addressed in the conventional heating algorithm in future experimental runs.

Implementing this smart thermostat methodology in a real building could prove challenging because of various thermal loads that are present in a respective structure, such as internal and ventilation loads. Besides that, the variable weather conditions, such as solar irradiance and outdoor air temperature can also affect the heating requirements of a building. Not only that, but different climate zones and building types add variance to the possible behavior types that the smart thermostat will come up against. To face this challenge, further preprocessing of the data may potentially increase the accuracy of the model (via outlier removal or data normalization), which is also something to be considered in future modeling efforts of the operation of real buildings. In addition, multi-surrogate model structures could assist in capturing real building behavior across different seasons of the year, which have been proven a significant parameter in smart controlling of HVAC systems [26], while they could be beneficial in multi-zone or multi-consumer applications, such as DHNs [38]. In fact, multi-surrogate modeling has displayed its viability in seasonal DHN heating load prediction applications [41], with each sub-model mimicking the heating load behavior of the DHN in a distinct season. Moreover, multi-surrogate modeling has also been used in a spatial manner, i.e., each sub-model being able to predict the heating loads of a single consumer, leading to an aggregated heating load prediction approach [42]. In a similar fashion, this work could more systematically utilize the surrogate models in real buildings, where each one can capture different states and behavior of the buildings, such as those proposed in this work (i.e., heating and free-float states).

Moreover, as the results displayed, testing the model effectiveness on heating patterns that are the same as those on which it was trained would provide no additional value to this work. Instead, this is something that should be done in real building cases. By addressing the above, produced methodologies such as the proposed one have the potential to be scaled and adapted to real building cases, which are typically larger and more complex in structure than the environmental chamber. Not only that, but building cases could include building consumers of the local DHN, and even multi-zone buildings, such as offices or public service buildings.

In terms of the further refinement of smart thermostat capabilities in the future, the smart thermostat can enhance their own communication capabilities with the users by assessing their preferred comfort settings (e.g., indoor air temperature setpoints) by using information such as activity level, clothing, and user feedback on their perception of air temperature [43]. More specifically, the smart thermostat can include an adaptive comfort model that is trained directly from user feedback and available data from smart devices that occupants commonly wear or use (e.g., activity level or heart rate from smart watches) [44]. In case online learning is difficult to implement, offline trained model alternatives include the provision of representative occupancy schedules created by clustering collected smart thermostat data [32]. The development and implementation of such methodologies provide supplementary value to the effectiveness of a smart thermostat, i.e., its ability to optimally preplan the thermal schedule. On another note, the development of the proposed thermostat currently focuses on the aspect of the demand-side management. The potential use of a smart controller responsible for the supply-side management of the chamber's heating system would provide a good opportunity for developing a smart controlling methodology for energy producers, which is a core component of future DHNs [38].

5. Conclusions

All in all, this approach presents the process of developing and experimentally validating a smart thermostat, accompanied by the respective development and testing of its controlling strategies. The training of a surrogate model through the use of easily acquired operational data that are commonly accessible to all thermostats (heating signal and indoor air temperature) enables its implementation on an environmental chamber in order to optimize the thermal comfort requests from a user during real-time operation.

In the present case, the developed smart thermostat methodology is able to address the issue of achieving desired temperature levels during initial occupancy time intervals by implementing a smart preheating strategy in real-time indoor space operation, which is a presently discussed matter in the current literature. More specifically, the optimized operation is conducted according to maximizing occupant comfort, while using as little heating as possible. By utilizing a variable predictive horizon (a customized variant of the respective predictive horizon commonly found in MPC applications) in combination with surrogate models, it can continuously adjust its course of controlling strategy implementation planning with increasing accuracy when approaching the scheduling events. Therefore, the current setup is able to face limitations of predictive heating control applications such as predictive uncertainty and heavy computational requirements at the same time. Furthermore, the operation of the smart thermostat embedded model, namely the indoor space surrogate model can be independent from its training, which can take place either outside or inside of the thermostat environment and have low computational requirements, providing a low-cost way to address the limitations of thermostats existing in the current literature. It is worth noting that the implementation of the model in the real-time operation of the environmental chamber provides insight into potential issues the model faces that could manifest when trained on indoor conditions unaffected by the external environment, which are difficult to identify when applying the developed models in more complex building cases. It is also worth mentioning that the environmental chamber provides an indoor space of a size comparable to single-zone building rooms that are usually examined in simulation efforts in the literature.

The results of the testing phase show that the current methodology has a promising model training and usage process, which does not require a lot of data and maintains a generalizable model that can be used in any heating schedule. In addition, its implementation on a preheating controlling strategy can lead to maximized thermal comfort levels, as test results provided *TC* percentage in the range 97.2–100%. However, present heating turnoff delays can significantly reduce thermal comfort. Besides that, the use of multi-surrogate model structure appears to be effective in increasing model predictive capabilities. Another observation made from the results of the experiment is that a potential measurement drift has been identified in the training measurement of this work. This can currently be presumed to be attributed to either existing thermal inertia in the chamber or innate sensor drift, although it will be addressed in future efforts.

Despite the benefits the proposed smart thermostat experimental setup provides, there is room for improvement in the following aspects. First of all, the implementation of the experiments could consider the thermal inertia phenomena discussed. For instance, measurement periods could avoid being conducted in consecutive days, to ensure that the initial conditions of the environmental chamber remain the same, while wall temperature measurements could be conducted after each operation session. Additionally, effective model training requires the provision of high-quality measured data. While temperature can be measured through relatively inexpensive means, the accuracy and precision of measurements can affect the quality of the developed model and, by extent, its ability to optimize the thermal operations of the building. Therefore, it is important that the measurement and data collection processes (preprocessing) are established more systematically. Future efforts could also focus on collecting energy measurements related to the heating system operation of the environmental chamber, leading to the development of energy-saving thermostat strategies.

Moreover, the current experiment took place in a closed environmental chamber, which is a controlled environment. Implementing this smart thermostat methodology in a real building could prove challenging because of the various thermal loads that are present in a particular structure, such as internal and ventilation loads. Besides that, the variable weather conditions, such as solar irradiance and outdoor air temperature, can also affect the heating requirements of a building. Furthermore, the thermal load profiles can vary depending on the use of the building, its characteristics, and the climate zone it is located in. Employing multi-surrogate modeling may be the answer to validating building models in different climate zones, as well as adapting to related heating patterns and behaviors that are linked to the location of the building and the complex heat loads of large-scale thermal networks. This makes a future pilot study worth conducting, in which smart controllers embedded with various types of surrogate models will be tested in real building cases of the local DHN. Such a study will be able to provide insight into the next steps for further developing and validating the smart thermostat modeling methodology, while examining the performance of the produced smart thermostat models on multiple types of indoor thermal zones, variable external environment conditions and heating load patterns. Lastly, regarding the broader aspect of energy management, having a controller with communicative abilities towards both its users and energy suppliers will be of great significance to establishing an integrated demand-and-supply management structure, while still allowing for autonomy in the actions of consumers. Not only that, but lightweight controlling strategies are a core part of future urban-scale DHNs because they encourage the inclusion of RES producers and prosumers in thermal networks, consequently facing upcoming needs of future thermal networks. In any case, in order to maintain effectiveness, surrogate modeling techniques used in smart thermostats should emphasize maintaining low-cost data acquisition techniques in order to lead to the development of autonomous and widely affordable smart thermostats with satisfying accuracy that can participate in urban-scale energy management efforts.

Author Contributions: Conceptualization, L.Z., G.P. and N.P.; methodology, L.Z., N.T., N.P. and G.P.; software, N.P.; validation, L.Z. and N.P.; formal analysis, L.Z.; investigation, L.Z.; resources, G.P. and N.T.; data curation, L.Z.; writing—original draft preparation, L.Z.; writing—review and editing, N.T., N.P. and G.P.; visualization, L.Z.; supervision, G.P.; project administration, G.P. and

N.T.; funding acquisition, G.P. and N.T. All authors have read and agreed to the published version of the manuscript.

Funding: This research received no external funding.

Data Availability Statement: Dataset available on request from the authors.

Conflicts of Interest: The authors declare no conflicts of interest.

Abbreviations

The following abbreviations are used in this manuscript:

1R1C	1-Resistance, 1-Capacitance model
ALAMO	Automatic Learning of Algebraic Models
DHN	District Heating Network
HVAC	Heating, Ventilation, and Air Conditioning
MPC	Model Predictive Control
PLC	Programmable Logic Controller
RBC	Rule-Based Controlling
RES	Renewable Energy Source
RL	Reinforcement Learning

References

1. EU. Energy Statistics—An Overview. Available online: https://ec.europa.eu/eurostat/statistics-explained/index.php?title=Energy_statistics_-_an_overview (accessed on 25 October 2023).
2. U.S. Energy Information Administration. Monthly Energy Review—February 2025. Available online: https://www.eia.gov/totalenergy/data/monthly/ (accessed on 19 March 2025).
3. Zhou, X.; Huang, Z.; Scheuer, B.; Wang, H.; Zhou, G.; Liu, Y. High-Resolution Estimation of Building Energy Consumption at the City Level. *Energy* **2023**, *275*, 127476. [CrossRef]
4. *Directive (EU) 2018/2001*; Directive (EU) 2018/2001 of the European Parliament and the Council of 11 December 2018 on the Promotion of the Use of Energy from Renewable Sources. The European Parliament and the Council of the European Union: Strasbourg, France, 2018.
5. European Commission. Communication from the Commission to the European Parliament, the European Council, the Council, the European Economic and Social Committee and the Committee of the Regions. The European Green Deal. Available online: https://eur-lex.europa.eu/resource.html?uri=cellar:b828d165-1c22-11ea-8c1f-01aa75ed71a1.0002.02/DOC_1&format=PDF (accessed on 3 June 2025).
6. European Commission. Stakeholder Consultation on the Renovation Wave Initiative—A Synthesis Report. Available online: https://energy.ec.europa.eu/system/files/2020-10/stakeholder_consultation_on_the_renovation_wave_initiative_0.pdf (accessed on 2 June 2025).
7. European Commission. Communication from the Commission to the European Parliament, the Council, the European Economic and Social Committee and the Committee of the Regions. A Renovation Wave for Europe—Greening Our Buildings, Creating Jobs, Improving Lives. Available online: https://energy.ec.europa.eu/system/files/2020-10/eu_renovation_wave_strategy_0.pdf (accessed on 3 June 2025).
8. Guelpa, E.; Verda, V. Demand Response and Other Demand Side Management Techniques for District Heating: A Review. *Energy* **2021**, *219*, 119440. [CrossRef]
9. Péan, T.Q.; Salom, J.; Costa-Castelló, R. Review of Control Strategies for Improving the Energy Flexibility Provided by Heat Pump Systems in Buildings. *J. Process Control* **2019**, *74*, 35–49. [CrossRef]
10. Liu, J.; Li, Y.; Ma, Y.; Qin, R.; Meng, X.; Wu, J. Coordinated Energy Management for Integrated Energy System Incorporating Multiple Flexibility Measures of Supply and Demand Sides: A Deep Reinforcement Learning Approach. *Energy Convers. Manag.* **2023**, *297*, 117728. [CrossRef]
11. Craparo, E.M.; Sprague, J.G. Integrated Supply- and Demand-Side Energy Management for Expeditionary Environmental Control. *Appl. Energy* **2019**, *233–234*, 352–366. [CrossRef]
12. Logenthiran, T.; Srinivasan, D.; Vanessa, K.W.M. Demand Side Management of Smart Grid: Load Shifting and Incentives. *J. Renew. Sustain. Energy* **2014**, *6*, 033136. [CrossRef]
13. Li, Y.; Wang, C.; Li, G.; Wang, J.; Zhao, D.; Chen, C. Improving Operational Flexibility of Integrated Energy System with Uncertain Renewable Generations Considering Thermal Inertia of Buildings. *Energy Convers. Manag.* **2020**, *207*, 112526. [CrossRef]

14. Vandermeulen, A.; van der Heijde, B.; Helsen, L. Controlling District Heating and Cooling Networks to Unlock Flexibility: A Review. *Energy* **2018**, *151*, 103–115. [CrossRef]
15. Bessa, R.; Moreira, C.; Silva, B.; Matos, M. Handling Renewable Energy Variability and Uncertainty in Power Systems Operation. *WIREs Energy Environ.* **2014**, *3*, 156–178. [CrossRef]
16. Martinez, S.; Vellei, M.; Dréau, J.L. Demand-Side Flexibility in a Residential District: What Are the Main Sources of Uncertainty? *Energy Build.* **2021**, *255*, 111595. [CrossRef]
17. Luo, M.; Zheng, Q.; Zhao, Y.; Zhao, F.; Zhou, X. Developing Occupant-Centric Smart Home Thermostats with Energy-Saving and Comfort-Improving Goals. *Energy Build.* **2023**, *299*, 113579. [CrossRef]
18. Zouloumis, L.; Ploskas, N.; Panaras, G. Quantifying Flexibility Potential on District Heating Local Thermal Substations. *Sustain. Energy Grids Netw.* **2023**, *35*, 101135. [CrossRef]
19. Magni, M.; Ochs, F.; de Vries, S.; Maccarini, A.; Sigg, F. Detailed Cross Comparison of Building Energy Simulation Tools Results Using a Reference Office Building as a Case Study. *Energy Build.* **2021**, *250*, 111260. [CrossRef]
20. Guo, F.; Rasmussen, B. Predictive Maintenance for Residential Air Conditioning Systems with Smart Thermostat Data Using Modified Mann-Kendall Tests. *Appl. Therm. Eng.* **2023**, *222*, 119955. [CrossRef]
21. Jiang, Y.; Andrew Ejenakevwe, K.; Wang, J.; Tang, C.Y.; Song, L. Development, Implementation, and Impact Analysis of Model Predictive Control-Based Optimal Precooling Using Smart Home Thermostats. *Energy Build.* **2024**, *303*, 113790. [CrossRef]
22. Zouloumis, L.; Panaras, G. Development of a Smart Thermostat. In Proceedings of the IAQ 2020: Indoor Environmental Quality Performance Approaches, Athens, Greece, 4–6 May 2022.
23. Arias, J.; Khan, A.A.; Rodriguez-Uría, J.; Sama, M. Analysis of Smart Thermostat Thermal Models for Residential Building. *Appl. Math. Model.* **2022**, *110*, 241–261. [CrossRef]
24. Vallianos, C.; Abtahi, M.; Athienitis, A.; Delcroix, B.; Rueda, L. Online Model-Based Predictive Control with Smart Thermostats: Application to an Experimental House in Québec. *J. Build. Perform. Simul.* **2024**, *17*, 94–110. [CrossRef]
25. De Coninck, R.; Baetens, R.; Saelens, D.; Woyle, A.; Helsen, L. Rule-Based Demand-Side Management of Domestic Hot Water Production with Heat Pumps in Zero Energy Neighbourhoods. *J. Build. Perform. Simul.* **2014**, *7*, 271–288. [CrossRef]
26. Safdari, M.; Janaideh, M.A.; Siddiqui, K.; Aliabadi, A.A. Weather-Adaptive Fuzzy Control of Setpoints for Energy-Efficient HVAC in Urban Buildings. *J. Build. Eng.* **2025**, *104*, 112317. [CrossRef]
27. Cho, S.; Park, C.S. Rule Reduction for Control of a Building Cooling System Using Explainable AI. *J. Build. Perform. Simul.* **2022**, *15*, 832–847. [CrossRef]
28. Queipo, N.V.; Haftka, R.T.; Shyy, W.; Goel, T.; Vaidyanathan, R.; Tucker, K.P. Surrogate-Based Analysis and Optimization. *Prog. Aerosp. Sci.* **2005**, *41*, 1–28. [CrossRef]
29. Lv, Z.; Niu, D.; Li, S.; Sun, H. Multi-Surrogate Assisted PSO with Adaptive Speciation for Expensive Multimodal Multi-Objective Optimization. *Appl. Soft Comput.* **2023**, *147*, 110724. [CrossRef]
30. Zhu, J.; Tian, Z.; Niu, J.; Lu, Y.; Cheng, B.; Zhou, H. Machine Learning-Enhanced Lightweight Rule-Based Control Strategy for Building Energy Demand Response. *Build. Simul.* **2025**. [CrossRef]
31. Stopps, H.; Touchie, M.F. Residential Smart Thermostat Use: An Exploration of Thermostat Programming, Environmental Attitudes, and the Influence of Smart Controls on Energy Savings. *Energy Build.* **2021**, *238*, 110834. [CrossRef]
32. Jung, W.; Wang, Z.; Hong, T.; Jazizadeh, F. Smart Thermostat Data-Driven U.S. Residential Occupancy Schedules and Development of a U.S. Residential Occupancy Schedule Simulator. *Build. Environ.* **2023**, *243*, 110628. [CrossRef]
33. Vallianos, C.; Candanedo, J.; Athienitis, A. Thermal Modeling for Control Applications of 60,000 Homes in North America Using Smart Thermostat Data. *Energy Build.* **2024**, *303*, 113811. [CrossRef]
34. Doma, A.; Prajapati, S.N.; Ouf, M.M. Developing a Residential Occupancy Schedule Generator Based on Smart Thermostat Data. *Build. Environ.* **2024**, *261*, 111713. [CrossRef]
35. Hou, D.; Allan, L.; Awad, H.; Bahiraei, F.; Evins, R. Estimating the Time Constant Using Smart Thermostat Data Acquisition and Manipulation: A Whole Building Experimental Study. *J. Build. Eng.* **2025**, *105*, 112485. [CrossRef]
36. Sun, S.; Xing, X.; Wang, J.; Sun, X.; Zhao, C. Preheating Time Estimation in Intermittent Heating with Hot-Water Radiators by Considering Model Uncertainties. *Build. Environ.* **2022**, *226*, 109734. [CrossRef]
37. Sun, S.; Wang, J.; Li, R.; Sun, Q. Estimation of Preheating Time for Building Intermittent Heating Subject to Changes in Outdoor Temperature and Solar Radiation. *Energy Build.* **2024**, *317*, 114405. [CrossRef]
38. Lund, H.; Østergaard, P.A.; Nielsen, T.B.; Werner, S.; Thorsen, J.E.; Gudmundsson, O.; Arabkoohsar, A.; Mathiesen, B.V. Perspectives on Fourth and Fifth Generation District Heating. *Energy* **2021**, *227*, 120520. [CrossRef]
39. Wilson, Z.T.; Sahinidis, N.V. The ALAMO Approach to Machine Learning. *Comput. Chem. Eng.* **2017**, *106*, 785–795. [CrossRef]
40. Gam, K.S.; Yang, I.; Kim, Y.-G. Thermal Hysteresis in Thin-Film Platinum Resistance Thermometers. *Int. J. Thermophys.* **2011**, *32*, 2388–2396. [CrossRef]
41. Kurek, T.; Bielecki, A.; Świrski, K.; Wojdan, K.; Guzek, M.; Białek, J.; Brzozowski, R.; Serafin, R. Heat Demand Forecasting Algorithm for a Warsaw District Heating Network. *Energy* **2021**, *217*, 119347. [CrossRef]

42. Wang, Y.; Li, Z.; Liu, J.; Zhao, Y.; Sun, S. A Novel Combined Model for Heat Load Prediction in District Heating Systems. *Appl. Therm. Eng.* **2023**, *227*, 120372. [CrossRef]
43. *ANSI/ASHRAE Standard 55-2017*; Thermal Environmental Conditions for Human Occupancy. ASHRAE: Peachtree Corners, GA, USA; American National Standards Institute: Washington, DC, USA, 2017.
44. Li, D.; Menassa, C.C.; Kamat, V.R. Personalized Human Comfort in Indoor Building Environments under Diverse Conditioning Modes. *Build. Environ.* **2017**, *126*, 304–317. [CrossRef]